Choosing Ecological Water Supply and Treatment

Judith Thornton

© Judith Thornton
September 2013
The Centre for Alternative Technology
Machynlleth, Powys, SY20 9AZ, UK
Tel. 01654 705950 • Fax. 01654 702782
info@cat.org.uk • www.cat.org.uk

Editor: Allan Shepherd
Illustrations: Graham Preston, Paul Bullen, Annika Faircloth, Judith Thornton

ISBN 978-1-902175-68-3

1 2 3 4 5 6 7 8 9 10

The details are provided in good faith and believed to be correct at the time of writing.
However, no responsibility is taken for any errors.
Our publications are updated regularly;
please let us know of any amendments or additions
which you think may be useful for future editions.

Printed and bound by CPI Group (UK) Ltd, Croydon, CR0 4YY

About the author

Judith Thornton is a post doctoral research fellow at the University of Leeds. Her research is in the area of water accounting and footprinting, water efficiency and water resilience. Judith has worked with the Centre for Alternative Technology since 1999, including as a Senior Lecturer within the Graduate School of the Environment, and as Head of Biology. She has installed a variety of water and sewage systems, including reed beds, compost toilets and private water supplies.

Contents

Introduction	1
The hydrological cycle – the way water flows	4
How much water is there and where is it?	4
Where did water come from originally?	6
The chemistry of water	7
How much do we need?	7
The taste of water: what's in it and what shouldn't be	7
Water and the environment	8
First steps	9
Further reading and websites	9
Chapter 1. Water use	11
Why should I use less?	11
How much water do we use and where does it all go?	15
How to use less water	19
Water demand for a house on a private water supply	29
Virtual water and the water footprint	30
Summary	31
Further reading and websites	31
Chapter 2. The mains and the alternatives	33
Water supplies and sustainability	34
Finding a private water supply – assessing the site	37
Getting the water out	45
Source protection	55
Summary	56
Further reading and websites	56
Chapter 3. Water treatment – underlying theory and regulation	59
What do we want to achieve by cleaning the water?	
What quality of water do you require?	60
Regulations on private water supplies in the UK	62
Microbiology and water treatment	72
If my water is so dangerous, why am I not ill yet?	73
What next?	74
Summary	75
Further reading	75

Chapter 4. Cleaning water – practicalities   77
Step 1 – removing solids   78
Step 2 – removing or killing the living organisms   84
Step 3 – controlling commonly occurring dissolved substances   89
Treatment techniques for dissolved materials   94
Standards for water treatment techniques   100
When to clean the water   100
Mains water supplies   101
Bottled water   106
Summary   106
Further reading and websites   107
Chapter 5. Fitting it all together   109
Water storage   109
How much energy do you need to move your water?   112
Pipes   116
Pumps   118
Other components   129
Contingency planning   130
Final touches – commissioning, inspection and record keeping   134
Summary   135
Further reading   135
Chapter 6. Rainwater harvesting   137
What can rainwater be used for?   137
How to use rainwater   138
Sizing the storage tank   148
Maintenance, monitoring and reliability issues   153
Environmental considerations surrounding RWH   154
Using RWH to meet water requirements under the
Code for Sustainable Homes (CSH)   156
Will a domestic rainwater harvesting system save me money?   156
Summary   158
Further reading and websites   158
Chapter 7. Grey water recycling systems   161
Which water sources are worth recycling?   162
Methods for recycling grey water   164
Does domestic grey water recycling have a future in the UK?   165
Variations on grey water recycling systems   166
Summary   167
Further reading   168

Chapter 8. Water in the garden   169
  Avoiding the need for watering   170
  Using water efficiently in the garden   173
  Regulations that apply to water in the garden   187
  Summary   188
  Further reading and websites   189
Conclusions   191
Appendix 1 – Private Water Supply Regulations   193
  Risk assessment   194
  Monitoring requirements   194
  Action in the event of water samples failing standards   196
  Differences between England, Scotland, Wales and N. Ireland 197
Appendix 2 – Checklist of actions   199
Acknowledgements   207
Glossary   209
Index   213

Introduction

Are you thinking about installing your own water supply? Do you already have one and want to understand more about it? Or do you want to build a house in an isolated area where the availability of water may well determine whether the project is feasible or not?

At the Centre for Alternative Technology, we look at all resources (including water) from the viewpoint of sustainability. This book therefore looks at the environmental, social and economic impacts of supplying and treating water from various sources. Whilst it is technically possible to treat almost any water to almost any desired standard, it does not necessarily follow that this is an environmentally friendly or appropriate thing to do, particularly if a water supply infrastructure already exists.

Consequently, this book does not advocate self-sufficiency for it's own sake, and 'closed loop' systems, in which dirty water that has already been used is recycled, are considered in light of their energy and resource requirements. Some would argue that energy intensive processes for water purification are acceptable if powered by renewable electricity, but this is like suggesting that leaving your lights on all the time is acceptable if you generate your own electricity; renewable energy should not be used as justification for profligate energy use!

Regardless of whether we are using a private water supply or the mains, all of us should concern ourselves with using less water. Increases in environmental awareness, combined with increasingly frequent hosepipe bans and drought orders, have also stimulated interest in systems that supplement mains water with water from other sources such as rain.

If you are going to rely on your own water supply, you need to know where to look for water. This introduction therefore considers the hydrological cycle – where water is on the planet and how it moves from place to place. It also describes some of the basic properties of water; these provide a context to the issues surrounding moving and cleaning it.

You may have to go to considerable trouble to find a water source and to clean it; it therefore makes sense not to use more of it than you need. **Chapter One** will help you calculate your water demand so that you can consider potential supplies in the light of your needs. The water efficiency measures described also provide a good starting point for those of you who have a mains water supply, and are interested in the environmental benefits of using less water.

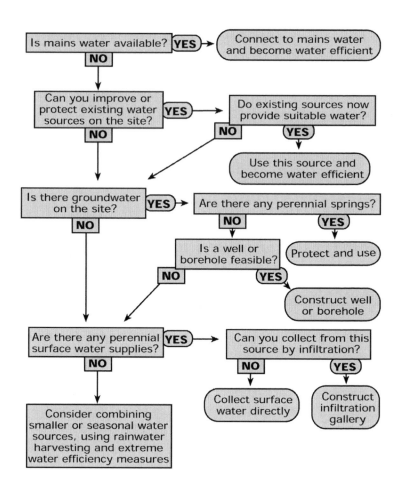

Figure 0.1. The decision flow chart.

Armed with a basic knowledge of the hydrological cycle and your water demand, **Chapter Two** will help you examine in detail how to find water. There will often be more than one source available to you, and you will need to consider the advantages and disadvantages of each. An overview of this decision making process is shown in the diagram above; the rationale behind it will become clear as you read through the chapters.

Once you have found your water source, you need to think about: how to look after it; what the water quality is compared to what you need it to be; and what might affect its quality. These issues are dealt with in **Chapter Three**. This chapter also introduces the legislation on private water supplies. Significant differences in regulations governing water supplies in Scotland, Northern Ireland or Wales (compared to those in England) are indicated.

How you clean water will depend upon what needs removing, and details of cleaning techniques are in **Chapter Four**. The next stage is to think about how to fit source, treatment and storage together, and get water in to your house. This is dealt with in **Chapter Five**. **Chapter Six** deals specifically with rain water as a source, and **Chapter Seven** considers grey water recycling, both of which are additional water sources that you may be looking at if water is in short supply. **Chapter Eight** considers the use of water in the garden – this can make up a considerable proportion of household water use in the summer. These three chapters are designed to stand alone and will be of interest to those who already have a mains water supply but are looking to supplement it.

Case studies are provided throughout the book to illustrate specific points, and to provide examples of how the principles can be translated into practice. Suggestions for further reading related to the content are given at the end of each chapter.

The book is predominantly aimed at the situation in the UK; solutions appropriate in other countries may not be applicable here for various reasons including water resource distribution and availability, politics, economics and social factors. This is true particularly of the more drastic water supply and recycling options that are the norm in areas where water is scarce or where mains water is less widespread; if the mains water supply pipe is tens of miles away a totally different system will be appropriate compared to in a household that already has mains water.

This book does not consider in any detail either the quality of tap water or the global politics surrounding water supply. Nor is this book about the processes of cleaning water after it has become sewage, for which the reader is referred to *Choosing Ecological Sewage Treatment.*

The hydrological cycle – the way water flows

Water is continually moving between the land, the sea and the air. The cycle in which it moves is called the hydrological cycle, and is illustrated in figure 0.2. The energy driving the cycle comes from the sun. Water in the sea and on the land surface evaporates and forms clouds. Movement of water from land to air is supplemented by transpiration – the process by which plants lose moisture to the atmosphere – so the hydrological cycle is closely linked to ecosystems. In the atmosphere, water forms clouds and these deposit their water back on the land as rain, snow or hail.

This water collects in streams and rivers and flows back into the sea. Some water soaks into the ground, and percolates through soil and rocks; this is known as groundwater, and it may return to the surface in lakes, streams or rivers, or it may return to the sea directly. This is the basic hydrological cycle, but you should bear in mind a few complications.

Firstly, there are a number of points at which water may miss out parts of the cycle, for example when it rains over the sea (this actually accounts for about 90% of total global precipitation). Secondly, the rate of water movement through the cycle varies. These variations can be both seasonal and geographical.

In tropical countries water may evaporate into the sky within minutes of falling as rain; conversely water can be trapped in ice caps for hundreds of thousands of years. Some groundwater is referred to as 'fossil water', since it has been underground for hundreds of thousands of years and would take similar times to be replenished if it was abstracted.

How much water is there and where is it?

The amount of water on the planet is constant, but is moving continually through the hydrological cycle. At any one time, the vast majority of water (97%) is in the sea. The table on page 6 indicates the total amount of water on the planet, and the amounts of water in each phase of the hydrological cycle at any time.

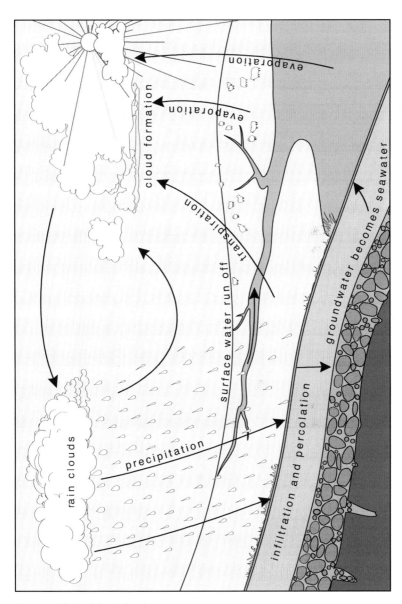

Figure 0.2. The hydrological cycle.

Location	Volume Tm^3	% total water
Sea	1320000	97.25
Polar ice caps, glaciers, snow	29200	2.1
Saline lakes and inland seas	105	0.008
Atmosphere	13	0.001
Groundwater	8250	
Soil moisture	65	0.62
Rivers	1.25	
Fresh water lakes and reservoirs	125	

Table 0.1. Distribution of water on the planet.
A note on units: One Tm^3 = 10^{12} m^3, and there are 1000 litres in a m^3 of water.

Whilst sea water accounts for the largest percentage of water on the planet, turning it into fresh water is an energy intensive process that is best suited to large scale (in other words thousands of households) applications in otherwise water scarce areas. It is certainly not an ecologically sound method to attempt on a domestic scale in the UK; we are more interested in groundwater, soil moisture, rivers and lakes. As you can see from the table, just 0.62% of the planet's water is available to us from these sources.

Where did water come from originally?

Water has been present in the solar system since the Big Bang, having originated during the thermonuclear fusion events that formed the original elements and compounds in the universe. Its presence on earth is as a result of water vapour being given off by the earth's mantle during volcanic eruptions and movement of the mantle several billion years ago. Volcanic eruptions continue to release minor quantities of water into the hydrological cycle today, although this probably amounts to only about a km^3 of water on the planet each year. Comets also contribute small amounts of 'new' water to the earth's hydrological cycle; they are made up largely of ice that vaporises and becomes atmospheric water as they near the earth.

The chemistry of water

Water (di-hydrogen monoxide if you want to make it sound dangerous) consists of two hydrogen atoms bonded to one oxygen atom: H_2O. It has some unusual properties owing to the existence of the hydrogen bond, which describes the tendency of one side of a water molecule to be slightly positively charged and the other side slightly negatively charged. This causes the water molecules to stick together, so water is liquid at ambient temperatures in the UK (unlike other compounds of similar molecular weight, which tend to be gases).

Water is also an excellent solvent, meaning that other molecules such as gases, salts and minerals are easily carried by it. Unfortunately, for us, this means water gets dirty very easily. Since this 'dirt' will include nutrients, there can be lots of micro-organisms in water, and some of these are harmful to us. In addition, the specific heat capacity of water is very high; put simply this means it takes a lot of energy to heat up water by small amounts. This is also advantageous to the creatures that live in water, as it makes it an easier and more constant environment to live in than air.

How much do we need?

Our bodies consist of about 60% water. We are continually losing this from our skin, through exhaling and as urine and faeces. Public health guidelines suggest drinking around 2-3 litres per day, but the amount you actually need varies with age, food intake, temperature, exercise and type of beverage. Development agencies regard 5 litres per person per day as the minimum for cooking and drinking purposes, rising to 15 litres if water is also used for bathing and clothes washing. If water-borne sanitation is required, this figure increases to 50 litres. As you will see in the next chapter, in the developed world most of us use far more than this.

The taste of water; what's in it and what shouldn't be

Totally pure water with nothing dissolved in it doesn't taste nice and isn't good for you; for example the calcium and magnesium carbonates that cause water hardness are useful minerals for bone growth. Cleaning water isn't always about taking 'bad' things out of it, the process can consist of putting things in.

Lots of things in water alter its flavour, often for the better, but in quite abstract ways. Water may taste 'flat' or 'energetic' according to the amount and type of dissolved gases in it. Sometimes water tastes almost sweet, sometimes sharp. The public water supply is regulated by the Drinking Water Inspectorate (DWI) and limits for over 50 parameters (including a range of chemicals, metals and bacteria) are checked.

We are exceptionally lucky in the UK that drinking water is so safe; several million tests are carried out on the public water supply every year and in 2010 99.96% passed. But even if it is safe, you may not like the taste of it. If you don't find the taste of tap water to your liking, the section in Chapter Five on additional treatments for public water supplies might help.

More and more people are drinking bottled water, despite the fact that it is often several thousand times more expensive than tap water. You may prefer the taste, but it is no better for your health than tap water (and for some people can be much worse). Bottled water is clearly worse in terms of its environmental cost; it may be pumped out of the ground at an unsustainable rate, put into a bottle whose production consumes more water than the bottle itself contains, and then transported half way across the world.

Water and the environment

We're not alone in needing water; it plays important physical, chemical and biological roles in the environment. These functions make water a vital part of all ecosystems.

Physical processes: At a local level, water is involved in warming and cooling via the processes of evaporation and condensation. At a global scale, the differences in density caused by temperature and salinity result in the thermohaline circulation which moves heat around the planet (with a net transfer of heat from the poles to the tropics). Water also has a 'lift up and carry away' function; carriage of silt and sediment by water is vital for the formation of river deltas, wetlands and flood plains. Water is also the largest contributory gas to the greenhouse effect, although its role is complicated because it has a rapid turnover in the atmosphere and the amount of water vapour is itself a function of temperature.

Chemical processes: As already stated, water is an excellent solvent. This means it plays a major transport role in many nutrient cycles (especially the nitrogen and phosphorous cycles) and is also an important medium for the spread of pollution.

Biological processes: Water and carbon dioxide are the building blocks for the production of sugar and oxygen in the process of photosynthesis. Water also plays other roles in plants, including transport, cooling (via transpiration) and creating structural integrity.

First steps

If you have recently moved to a house served by a private water supply and aren't sure where to start, or have decided to do something about an existing supply, then you may be a bit overwhelmed. Your first instinct may be to have a water sample from the kitchen tap tested to see what's in it, or to bring in a professional to tell you what to do.

However, many of the measures a professional will recommend are extremely simple so it is generally worth doing some preliminary investigations yourself using table 3.3 (on pages 65-68) as a starting point. Managing your water supply will require some effort on your part, so understanding the components of the system, including finding them on the ground and then drawing them onto a map is an obvious first step.

Table 3.3 will give you a list of jobs to do. These will generally be measures such as the source control techniques detailed in Chapter Two. Once you have done the simplest of these, getting a water sample tested as discussed on page 69 will be your next step. As you will see, it isn't always straightforward to decide what to get the sample tested for, but your local Environmental Health team should be able to offer (free) advice on this.

Once you have the test results, and have confirmed that there aren't any insurmountable supply issues, you can develop a more detailed plan of works based on the findings from table 3.3. As discussed on page 73, even apparently clean sources will need at least some treatment. Chapter Four discusses the treatment technologies that are available.

Further reading

Bottled water: Understanding a social phenomenon.
Available from www.wwf.org.uk

Choosing Ecological Sewage Treatment, Grant, Moodie and Weedon, CAT Publications, 2012. Available from CAT Ecostore.

Engineering Hydrology, Wilson, Palgrave Macmillan, 1990. Data for table 0.1.

Peak Water, Palaniappan & Gleick, 2009;
Chapter from *The Worlds Water*, Pacific Institute.

Websites
Drinking Water Inspectorate – UK government organisation responsible
for the quality of drinking water. Website includes factsheets on water
quality, and annual report on mains water quality. http://dwi.defra.gov.uk/
stakeholders/private-water-supplies/

Chapter One

Water use

How much water you use – why use less? – water use and behaviour – water using appliances – bottom line water consumption – virtual water

As you saw in the Introduction, there is a limited water resource available. If you have a mains water supply you should read this chapter and implement water efficiency measures before considering supplementing your water from other sources. In fact, if what you are interested in is your total environmental impact, you should bear in mind that for most people other aspects of lifestyle such as transport, food and energy consumption are much more environmentally damaging. Your 'virtual water' use (see later in the chapter) may also be more significant than your domestic water use.

Why should I use less?

Having had clean, cheap and plentiful water on tap for several generations, many people have little idea about why they should use less water, and we have a widely held assumption that the UK is 'wet'. In fact, the water exploitation index (the relationship between the amount of water available and the amount abstracted) means that the south east of England is classified as being under water stress. In addition to the general issue of water stress, there are a number of other reasons to consider:

- Reducing the impact of water use on greenhouse gas emissions.
- Decreasing the demand for potable (in other words drinking quality) water.
- Reducing demand for groundwater and the threat to rivers from over-abstraction.
- The effect of climate change on water in the UK.
- Improving sewage treatment.
- Reducing water and sewerage bills.

Reducing the impact of water use on greenhouse gas emissions:
The water that comes out of your tap will have been through a number
of processes to make it clean and safe. This can include screening and
filtering to remove solids, coagulation and flocculation to remove colour,
and disinfection, to inactivate pathogens. All of these processes have an
environmental impact and reducing water use will help minimise these
impacts. However, as shown in figure 1.1, these impacts are minor compared
to those from our use of hot water in the home.

Three things follow from this. Firstly, worrying about what the
water/sewage company is doing to reduce their environmental impact
is rather trivial compared to thinking about your own impact. Secondly,
'doing your bit' to reduce the environmental impact of your water use entails
using less water rather than installing water recycling systems. Thirdly, for
most people, your water related impact is a very small amount compared to
total household CO_2 emissions. If your primary concern is climate change
you should pay attention to other areas of your lifestyle before worrying
about water use.

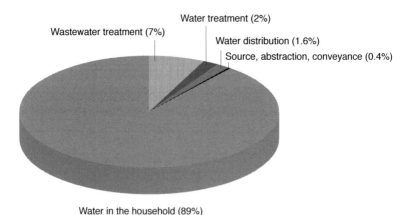

Figure 1.1. The CO_2 emissions within the water supply-use-disposal chain.
Based on a study by the Environment Agency (2008). Science Report SC0700010.

Decreasing the demand for potable water: Many areas where
populations are projected to increase are already water stressed. It stands
to reason that the cheapest method for meeting the water needs of the
increasing population is for all of us to use a bit less.

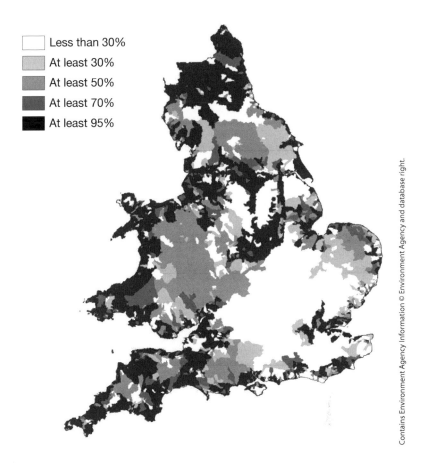

Less than 30%
At least 30%
At least 50%
At least 70%
At least 95%

Contains Environment Agency Information © Environment Agency and database right.

Figure 1.2. Water resource availability, based on Environment Agency catchment abstraction management strategy (CAMS) data from 2010. It indicates the percentage of the year during which additional water is available in the environment for abstraction and use.

Reducing the impacts of water abstraction on the environment: Aquatic habitats are very sensitive to the amount of water in them, so flow and water level variability must be protected. The Environment Agency's Catchment Abstraction Management Strategies (CAMS) are regionally specific and detail how this should be done in England and Wales. In many areas we are removing water from ground- and surface water at unsustainable rates already, and in others there is no additional water available (figure 1.2).

The effect of climate change on water availability in the UK: Predictions of the severity of climate change are full of uncertainties. However, regardless of the average temperature increase that will occur, there is little doubt that weather patterns in the UK will become more extreme. In terms of water, this will mean higher winter rainfall, increased flooding, and lower summer rainfall. This will increase the need for inter-seasonal water storage in large reservoirs and will necessitate a change in our approach to storm water drainage.

Improving sewage treatment: The water that you use in the house has to be treated before releasing it into the environment. You probably use around 50 litres of water each day just to flush the toilet. Other less dirty 'grey' water from showers, baths, washing machines and so on is added to the mix. Water is a convenient transport medium that can carry solids with it along pipes to a treatment works, but the treatment processes are generally designed to remove the solids from the water again. Consequently, using the bare minimum of water allows the cleaning process to be easier, cheaper and have a lower environmental impact.

Reducing water, energy and sewage bills: The number of households with water meters (currently around a third) is set to rise dramatically over the next 5-10 years (possibly to 80%). Paying for a resource according to how much you use of it is generally regarded as equitable as long as affordability safeguards for low income households exist. Water charges in the future may also reflect the time of year (for example charging more for water during times of summer scarcity), and the volume ('rising block tariffs' where a relatively low amount is charged to low water users, but progressively higher unit charges apply as more water is used over a specified threshold).

You can request a water meter from your water company; they will indicate whether they think you are likely to save money if you install one, and the meter will be installed free of charge. If it is impractical to install a meter (for example if you have a shared supply pipe), the water company can opt to base your bill on a survey based assessment of what your water use is. If you have a water meter, your sewage bill will be calculated in relation to your mains water usage (based on an assumed return of 95% of your mains water as sewage).

Being efficient with hot water will save you money on your energy bills as well as your water bill. The relative amounts of these savings vary according to how you heat your water, and which energy and water supplier you have, but the value of the energy saving can outweigh that of the water saving.

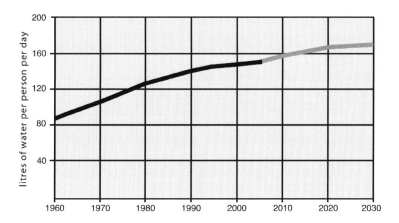

Figure 1.3. Household water consumption in the UK (courtesy of the Environment Agency).

How much water do we use and where does it all go?

Domestic water use per capita in the UK has risen from 100 litres per person per day (l/p/d) in 1970 to 150 l/p/d in 2008/9 (figure 1.3). This increase is due to a number of factors including smaller households (per capita use is higher in small households than larger ones), cultural shifts towards increased washing and bathing, and increasing affluence – resulting in higher numbers of washing machines, dishwashers and larger gardens.

In recent years water use has stabilised, the trend towards increased water use from bathing has been largely counterbalanced by improvements in the water efficiency of WC's. Most future projections of domestic water use show a slight downward trend, driven by increases in water metering and regulations on water efficiency. New build properties must comply with Part G of Building Regulations (Sanitation, hot water safety and water efficiency). This specifies a maximum limit of 125 l/p/d, calculated using the 'water calculator' (discussed further in the box on page 18). Since Building Regulations are a devolved responsibility, differences exist in Scotland, Wales and Northern Ireland. The domestic water use of the UK is compared to that of other countries in figure 1.4.

Average water consumption obviously hides a lot of individual differences. If you have a water meter it is easy to find out how much water you use, either from your bill or by reading the meter (figure 1.5). If you don't have a water meter, it's a little more difficult. Bizarrely, the water companies don't

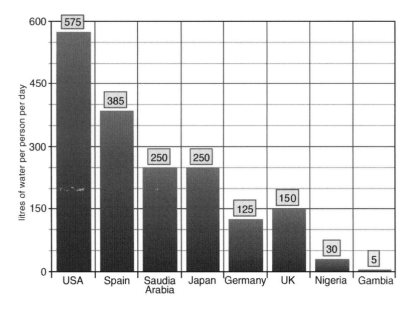

Figure 1.4. Household water consumption in the UK compared to other countries (courtesy of the Environment Agency).

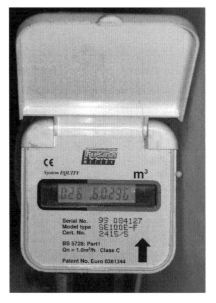

have accurate figures for how much is used within households. This is because whilst they know how much they supply into their distribution network, and they also have meters at district level, it's very difficult to calculate the amount that is lost to leakage during distribution.

Figure 1.5. Domestic water meter. Meters installed by water companies measure the volume of water used in m³. A cubic metre of water is 1000 litres, so since the installation of this water meter, the householders have used 26.6 x 1000 litres of water, i.e. 26,600 litres.

Box 1.1. Calculating your water use

The amount of water that an appliance uses varies according to its age and manufacturer. However, the biggest cause of variation is behaviour. For appliances that have a fixed use (for example toilet, washing machine, dishwasher), the water use is fairly similar each time you use it. This means you can simply multiply the volume per use by the number of times it is used. It can be much more difficult to calculate how much water is used by other appliances - such as taps, showers and the bath - since we interact with them in quite different ways.

For these appliances the flow rate can vary according to who is using it, and the durations of use can be very different as well. There is no good substitute for doing your own measurement for these appliances (buckets, measuring jugs and stopwatches will come in handy). This will enable you to get an idea of how much water you use, as well as how much it varies and for what reasons. If you're really keen, plumb in some flow meters to look at water use in the house. Some typical data is given in table 1.1.

Appliance	Volume or flow rate	Uses/day
Toilet	6.5 litres/flush	5
Shower	9 litres/min	1
Power shower	18 litres/min	
Bath	70 litres/use	0.2
Washing machine	50 litres/cycle	0.3
Dishwasher	18 litres/cycle	0.5
Kitchen taps	20 litres/day	
Washbasin	14 litres/day	

Table 1.1. Some approximate water use data for a 'normal' individual.

The regulator (OFWAT) estimates that total leakage during supply in 2008/2009 was 3650 mega litres per day (equivalent to 135 litres per property per day). This amount will decrease as ageing infrastructure is repaired, but the economic and environmental costs of digging up water supply pipes and replacing them with new ones limits the rate of potential improvements.

You can work out where your water goes by doing a water audit as shown in the box above. You may well find that simply by measuring your water use and thinking about the issues a little more you will naturally start using less water; carrying out an audit can be an excellent water efficiency measure itself.

It is also a good opportunity to learn more about the layout of the water system in your house; location of the stop tap (check it works!), location of any header tanks (and whether they are properly covered, or have dead things floating in them) and what appliances are fed from the header tanks. In many older houses, the cold water taps in the bathroom are fed from a header tank rather than direct from the cold water pipe coming in to the house; in this situation it is particularly important to be aware of the cleanliness or otherwise of the header tank.

Box 1.2. Water efficiency incentives and regulation

There is increasing emphasis on water efficiency at policy level, and this has led to a number of changes in regulation, incentive and labelling schemes. The cornerstone of government policy on this is the 'water calculator' in the Code for Sustainable Homes (CSH), which is also incorporated into Part G of Building Regulations. The CSH has mandatory water efficiency levels, alongside requirements for other environmental measures such as energy efficiency, waste, pollution and use of materials.

There are two basic approaches to water efficiency schemes; basing them on the individual water using appliances (a 'fittings based approach'), or on the total water consumption of the house (a 'whole house approach'). CSH uses the latter approach, ranging from maximum consumptions of 120 litres/person/day (l/p/d) at the lower code levels, down to 80 l/p/d at the highest code levels.

This approach has a number of problems, including the implied accuracy (calculations are to the nearest 10 millilitres per person per day), and the fact that it assumes set use frequencies by the householders; that we flush the toilet 4.42 times a day and have 0.11 baths a day, if we also possess a shower. Other disadvantages include lack of suitability for retrofit schemes or replacement of individual appliances, and the ability to offset one appliance against another, so that (for example) a power shower that is profligate in both water and energy terms can be offset by recycling rainwater for toilet flushing (which is itself environmentally unsound (as discussed in Chapter Six)).

The contrasting 'fittings based approach', has a number of advantages and it fits well with the Water Efficiency Product Labelling Scheme (WEPLS), which rates appliances according to their water use, using a similar format to the energy labels common on white goods (figure 1.6). Scottish Building Regulations are more enlightened than their English counterparts, and use a fittings based approach relating to WEPLS. These issues are discussed further in the AECB Water Standard Technical Guidance (see further reading section for references).

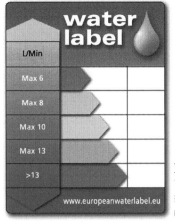

Figure 1.6. A typical label (for taps) from the Water Efficiency Product Labelling Scheme.

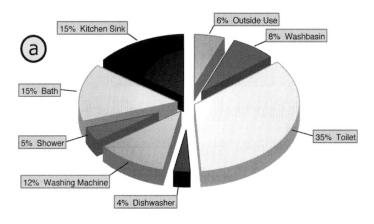

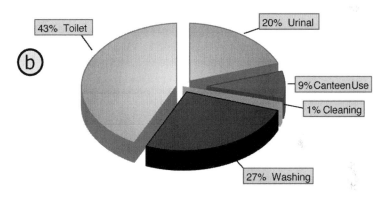

Figure 1.7a and b. Usage of water in a typical house (a), and office (b).
(Data courtesy of the Environment Agency and the Building Research Establishment).

How to use less water

An audit will help to determine where the biggest water savings can be made. The average domestic situation is shown in figure 1.7a. If you are interested in minimising water use in your work place, then figure 1.7b shows the typical water uses in an office. More detailed data on commercial water uses within specific sectors is available via the 'Watermark' project (details are listed in the Further reading section at the end of this chapter). There are two basic approaches to using less water. The first is to change your behaviour; the second is to change your water-using appliances. If you are building a new house, replumbing an old one or replacing your boiler, there may be more measures you can take, as discussed in the box *Plumbing system design* on page 28.

Behaviour

The following behavioural changes are fairly easy (and free!), and will result in substantial water savings. Savings involving hot water should be prioritised if you are interested in your household greenhouse gas emissions as well as your water use:

- Between 5 and 20 litres of water come out of a tap in a minute (depending on water pressure). Don't leave the tap on whilst brushing your teeth, washing and so on. You could easily be wasting 15,000 litres of water a year just brushing your teeth!
- Think about how you wash your hands. Waiting for the hot water tap to run hot wastes several litres of water each time; can you use this cool water for something else? Or if you don't generally bother waiting for the tap to run hot, use the cold tap instead; using the hot water just moves a few litres of hot water along a dead leg in the plumbing (see page 29 for explanation of dead legs).
- Use a washing up bowl rather than the kitchen sink. It only needs half as much water to fill.
- Wait for a full load before using the washing machine or dishwasher. The 'half load' or 'economy' settings use at least 75% of the energy and water compared to a full load, so it's not a 'half' load at all.
- Don't cool food under a running tap; measure the flow rate of your kitchen tap to work out how much water this wastes.
- Not flushing the toilet after urinating may be acceptable in some households, even better is to capture the nutrients in urine by saving it for the garden.
- Measure how much water your shower uses. Switch it off whilst you're using the shampoo and soap.

There are also lots of behavioural changes you can use to make your garden more water efficient and these are discussed in Chapter Eight. There are many other behavioural changes that may be applicable to your household over and above those listed here.

In a more abstract sense, it is worth considering our activities when using water; it is often part of habitual behaviours and daily routines rather than an actual 'need' for water. Showering is a good example; this often has little to do with being physically clean, but may simply be part of a routine such as waking up, relaxing, having time to ourselves.

People's morning routines are often deeply embedded, automatic and wasteful; shaving or brushing your teeth in the shower for example. When looked at dispassionately, in terms of the duration and flow rate of water, both these activities could clearly be undertaken with far lower water use. Understanding the importance of these habits and routines surrounding water use is vital in order to have sensitive conversations with the rest of your household about water efficiency.

How much does metering reduce water use?
Most UK studies have demonstrated that households on water meters use less water than un-metered households (anywhere from around 5% to 25%). However, since low water use households are more likely to request a water meter than high water use ones (and water companies will suggest whether or not a customer is likely to save by being on a water meter), this statistic cannot be applied more generally.

Compulsory metering is going to increase over the next 5 years, and it remains to be seen what the overall water saving will be. Availability of social tariffs will need to increase to avoid 'water poverty' (defined as when a household spends more than 3% of their net income on water and sewage charges).

It has also been argued that commoditizing water and the mentality of 'paying for what you use' may cause a minority of people to increase their water use as they then feel they have a right to. As 'smart meters' become more common for utilities in general, it is likely that approaches to water metering as a means of encouraging behavioural change will be modified (for example for the meter to be visible, and for hot and cold water to be considered separately).

Modifying water-using appliances
Once you're a master of water conscious behaviour, you can look at some technical fixes to reduce your water use. An increasing number of water using appliances now carry labels telling you how water efficient they are (figure 1.6). The AECB (the Sustainable Building Association) have also produced a water standard, with suggested maximum water uses for individual appliances (table 1.2).

Fitting	Good practice	Best practice
Showers	6 to 8 l/min	≤ 6 l/min (for example aerating)
Basin and bidet taps (domestic)	4 to 6 l/min	≤ 4 l/min with lower default
Urinals (non-domestic)	See AECB standard for details	
Basin taps (washroom)	≤ 1.7 l/min spray Dead leg ≤ 0.5 litres	As Good Practice Dead leg ≤ 0.25 litres
Kitchen sink taps	6 to 8 l/min	≤ 6 l/min with lower default
White goods	See AECB energy standard	As Good Practice
Toilets	≤ 6 l/ full flush	≤ 4.5 l/full flush
Baths (shower must also be installed)	≤ 180 litres to overflow	As Good Practice
Dead legs	≤ 1.5 litres	≤ 0.85 litres
Dead legs off secondary circulation	≤ 0.5 litres	≤ 0.25 litres
Water softeners	Location specific, see detailed AECB standard	
Outdoor	Location specific, see detailed AECB standard	

Table 1.2. The AECB Water Standards represent what the AECB (the Sustainable Building Association) regard as near optimum for water efficient appliances. See Further reading section for details.

Fixing leaks

Leaking taps or ball valves (such as in a header tank or toilet cistern) can waste enormous amounts of water. A slight drip wastes 30 litres a day. By the time the drips have built up to a trickle, it could easily be 300 litres a day. Both tap washers and ball valve washers are cheap and easily fixed (if you don't know how, refer to a DIY manual or youtube; no need to send for a plumber).

Toilets

As you saw from the pie chart in figure 1.7, a third of our domestic water is flushed down the toilet. Depending on how old your toilet is it may use up to 12 litres per flush. There are a number of things you can do to your existing toilet to make it more efficient.

A. Put a cistern displacement device in the cistern (figure 1.8a). Toilet cisterns marked 2 gallons (9 litres) on the inside of the cistern can usually flush the bowl with less water. You can either put a plastic bottle of water in the cistern, or get cistern displacement devices (often called 'hippos', 'hogs' or 'sava flushes') from your water company. Make sure your toilet still flushes first time with the reduced volume; if you have to flush twice you will be wasting water and should remove the cistern displacement device or try replacing it with a smaller one. If you have a brick in your toilet cistern, it's best to remove it and replace it with one of these devices, since over time the brick will crumble and damage the flush mechanism.

B. Modify the flush lever. Retrofit a variable flush lever to your cistern (figure 1.8b). These interrupt the siphon during the flush cycle, so the water flow stops. Since they operate differently from conventional flush levers you should make sure that you put written instructions next to the toilet so that people don't waste water by using the wrong setting and flushing more than once.

C. Install a delayed opening float valve. Normally, water starts to refill the cistern whilst it is flushing, and this waste a litre of water every flush. Delayed action inlet valves have a small chamber underneath the valve that delay the cistern refilling until after the flush has finished (figure 1.8c).

Other modifications to the flushing mechanism on older toilets are possible (for example various types of interruptible flush in which the flush volume varies according to the duration the flush lever is held down for). However, many of these have been found to result in 'double flushing', so create increased water use unless there is very clear signage about how the flush mechanism works and you are willing to educate the users.

The above measures will help reduce water consumption in your existing toilet, but if you are considering buying a new one, the right choice will reduce flush volume to about four litres. A variety of models are available and there are two basic mechanisms available; siphon flush and valve flush. Valve flush toilets can flush with less water than their siphon containing counterparts, and are more likely to have a 'half-flush' button, which may result in further water saving (figure 1.9).

However, concerns remain about the long-term efficiency of valve flush toilets, as they are more prone to leakage than siphon toilets. Even if this is a rare occurrence, it can easily exceed the savings accrued from a lower flush volume. If you are the kind of person who will spot a leak and do something about it, get a valve flush toilet, otherwise you should stick to a siphon flush.

Figure 1.8b. Variable flush lever. Flush volume can be changed between 'min', 'med' or 'max' by twisting the dial on the front. This interrupts the siphon and hence changes how much water is used per flush.

Figure 1.8a. A cistern displacement device. These take up space in the toilet cistern so that there is less space available for water, thereby reducing the flush volume.

Figure 1.8c. Delayed action inlet valve. The chamber underneath the float valve contains water which drains more slowly than the toilet cistern. This prevents the float valve from opening until later in the flush cycle, thereby preventing water wastage (courtesy of Opella).

Figure 1.9. Dual flush toilet.

Vacuum flush toilets, as seen on planes, boats and trains are very water efficient, using around 0.5 litres per flush. In most systems the vacuum is continuous, that is to say that the pressure in the pipework is below atmospheric pressure. When the flush button is pressed, a valve opens in the toilet and the contents of the bowl are sucked along the pipe.

Clearly it takes energy to generate this vacuum, so this type of toilet is best suited to applications where water use issues are more important than energy issues (for example on aeroplanes, where the energy implications of a vacuum flush are minimal compared to the implications of carrying the volumes of water required for a full flush).

Lower energy vacuum flush toilets (in which the vacuum is only generated when the flush button is pressed, rather than being continuous) are available, and the usual reason for installing them domestically is where space or gradient considerations make sewer pipe runs difficult. For most of us though, energy is generally a bigger environmental concern than water.

The most water efficient toilet of them all is a compost toilet, where no flush water or energy is required and you have the added benefit of being able to recover the nutrients and use them in the garden. It is possible to fit a clean, sanitary compost toilet into a domestic bathroom.

Urinals

Urinals are unusual in the domestic bathroom, but if you have men in the household, and enough space in the bathroom, it may be worth considering. Waterless urinals will drastically cut water use, although check before you buy, because many have disposable cartridges, the production and disposal of which may have a higher environmental impact than the water saved. Smells from urinals are generally related to bowl or wall design issues that make cleaning difficult rather than water use, so are not inevitable.

Showers

Our shower habits are often as much to do with the sensation and experience of showering than about getting clean, so what suits one household will not necessarily suit another.

Shower flow rates vary depending on how the water is heated. The flow rate in an electric shower varies according to the temperature of the mains water it is heating, and is often less than 6 litres per minute. Flow rates in mixer showers are much higher, often up to 20 litres a minute. Whilst electric showers are inherently water efficient, they are less good from the perspective of CO_2 emissions (owing to the amount of fossil fuels used to produce electricity, known as the 'fuel factor').

Mixer shower flow rates can usually be reduced to around 6 litres per minute and still give good performance. Some shower heads do this by atomising the flow (delivering it in smaller droplets), others by aerating the spray (giving the sensation of greater volume), or by having a reduced number of jets. Flow restrictors also exist that can prevent the delivery of more than a set number of litres per minute, and may be available free from your water company. If you have a pumped shower, the solution is likely to be more complex and you should consult a plumber.

Baths

Standard bath volumes are around 200 litres (measured to the level of the overflow) and a typical bath uses around 70 litres. For most people, deciding between having a shower and a bath is not primarily related to water efficiency concerns; an occasional hot soak might make little difference to water use compared to a change in shower duration.

Taps

Between 5 and 20 litres of water come out of a bathroom tap per minute and up to 30 in the kitchen (the maximum flow rate is dependent on water pressure and the model of tap, but most of us rarely turn the tap on full anyway). The simple behavioural change of turning a tap off will result in substantial water savings, but there are also technological fixes that will reduce water use by taps:

- **Flow regulators** (figure 1.11). These are plumbed into the pipework leading to the tap and will limit the water to a maximum flow rate. They have the added benefits of reducing noise, splashing and water hammer and may help balance the flows within your house if several appliances are being used simultaneously.
- **Shut off mechanisms**. Various mechanisms are available to make taps turn off automatically. These include percussion taps (which you push to turn on, but turn off automatically) and taps with infra-red proximity sensors. Choose a model that allows you to adjust the length of time before the tap turns off.
- **Spray fittings**. The continuous flow of water coming out of taps isn't actually a very efficient way of getting wet! Spray fittings can reduce the

Figure 1.10. Tapmagic devices. These screw into the outlet from your tap. If you turn the tap on half way a spray is produced, but if the tap is turned further a full flow is obtained.

Figure 1.11. Flow regulator (courtesy of Green Building Store).

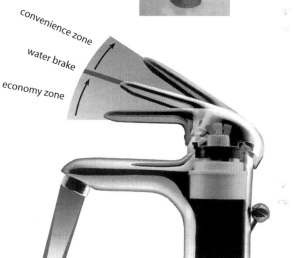

Figure 1.12. Tap with 'water brake'. On pulling the lever upwards, you have to overcome a noticeable resistance (called the 'water brake') to obtain a full flow rather then a low flow (courtesy of Hansa).

flow of water through a tap to less than 2 litres per minute, yet still result in adequate wetting. These are an excellent choice in situations where you don't need to fill the basin or sink (for example, for hand washing), but adjustable models are also available which allow a normal tap to be converted to one with two settings; both spray and full flow (figure 1.10). Check the minimum water pressure requirements of the model you are buying before purchase.

- **Taps with 'water brakes'.** These are lever taps in which the lever is easy to push to a certain point, and in this range the flow is low (for example under five litres per minute). To obtain a full flow you must push the lever further, past a noticeable resistance point (known as the water brake). The tap illustrated in figure 1.12 (by Hansa) also has an integral adjustable flow regulator.

Washing machines and dishwashers

The majority of modern washing machines use less than 50 litres per full load. From the perspective of CO_2 emissions, 'hot fill' washing machines offer theoretical advantages if a low carbon source of hot water is available (for example solar). However, because the volume of water used in the wash part of the cycle is small (about 17 litres of the total), the heat source needs to be very close to the washing machine to take advantage of this heat. Additionally, many machines only use the hot supply for washes above 60, so for most people there is unlikely to be any advantage in hot fill machines.

Modern dishwashers use 14 litres or less on a full load (of 12 place settings). But how does this compare to washing by hand? A standard washing up bowl filled to about an inch from the top holds approximately 7 litres, and a standard sink holds about 15 litres, but the way in which people do the washing up is so variable that there is no definitive answer in terms of water efficiency. In terms of energy efficiency, washing up by hand is preferable.

Garden

Whilst it does not account for a major proportion of our water use on an annual basis, use of water in the garden increases during the summer when demand is naturally high and supplies are most stressed. There are a number of ways in which you can garden in a more water efficient way, these are considered in detail in Chapter Eight.

Box 1.3. Plumbing system design

In most households, behavioural change and upgrading appliances is as far as it is sensible to go with water efficiency. However, if you are starting from scratch there is more scope for system level changes that can improve performance of water using appliances as well as saving water and energy.

Pressure regulation: mains water is generally supplied at between 1 and 3 bar (water pressure is discussed in detail in Chapter Five). High incoming water pressure often results in higher household water use, and a pressure regulating valve can be installed where the cold pipe enters the building in order to avoid this. The AECB standard recommends this where incoming water pressure is greater than 3.5 Bar.

Pipe diameters, dead legs and boiler placement: A dead leg (also referred to simply as a leg) is the term usually used to refer to the volume of water between the hot water source and the appliance it is supplying; it is therefore a function of pipe diameter and pipe length. Water in the pipe gradually cools down to room temperature, so the presence of a large dead leg is noticed as the duration of time it takes the hot tap to run hot. Consequently, dead legs represent a performance problem, as well as wasting water and energy and potentially increasing the legionella risk.

Owing to the history of gravity fed plumbing systems in the UK, water pipe diameters are large (for example 22mm) in order to minimise the pressure drop along the pipe. More modern hot water systems have higher water pressure, so decreasing the need for such large pipe diameters. Plumbing design guidelines have failed to change to take account of this. Pipe diameters of 10 or 12mm may well be sufficient for hot taps on wash basins. All hot and cold water pipes should be insulated, and where running in parallel the cold pipe should be below the hot, so that any heat escaping from the hot pipe doesn't rise and warm the cold pipe, (or vice versa for coolth escaping from the cold pipe). The location of the hot water source within the house also plays an important role in minimising the necessary pipe lengths (and therefore dead legs), so should also be considered.

Visible water meters and sub metering: 'smart meters' are increasingly talked about in relation to electricity use, but the general principle of increasing householder awareness of resource use also applies to water. Putting a water meter somewhere visible so that you can easily keep track of your water use can help, and separate meters on hot and cold water will help you keep track of your energy use from hot water.

Water demand for a house on a private water supply

Implementing behavioural changes and installing water efficient appliances can reduce your water use from a UK average of 150 litres per person per day (l/p/d) to 80 l/p/d fairly easily. Further savings can be made if necessary, and households with compost toilets and who are dedicated to water efficiency can use less than 40 l/p/d. If you are going to rely on a private water supply, you will need to decide how much, or how little, water you are likely to use, and this will determine what water sources are realistic, how much storage you will need, how to treat the water and how big the pipes will need to be. The sustainability of the water supply itself will also determine how water efficient you need to be.

There may be instances where no special measures are necessary, since the water you use will be going back into the same catchment that you abstracted it from, and there is plenty there, but in other circumstances (for example if you are relying entirely on rainwater harvesting) you may need to be very frugal.

You may be happy to make significant lifestyle changes and survive on minimum water use, but if you are planning to expand or sell the property, house more people, or have regular guests, you cannot necessarily expect the same low level of water use by other people. Convenience can have a major effect on water use; if you have to go to much effort to get water you will probably use less of it (for example if you choose to pump water by hand rather than electrically).

How variable is your water demand likely to be? Will the washing machine be in use at the same time as the bath or shower? What are the consequences of your supply running out for a couple of hours or even several days? Do you need to know you'll always have plenty of water or might you be willing to forego the washing machine and garden watering for a few weeks in the summer?

Is your water supply likely to be reliable during long dry periods or are you going to have to store water for several months? Sometimes a relatively minor increase in infrastructure cost can substantially increase your storage capacity. You may also decide to have a modular storage and treatment system that will easily allow for increases in future demand.

If this is the case you may choose to take a risk and estimate a very low daily water consumption but be willing to increase storage capacity at short notice if you get your estimates wrong. In most cases the best way of deciding how much water you are going to need is to fit a flow meter on to your water supply and measure your water use over as long a time period as possible.

Virtual water and the water footprint

This book is about the water you use directly. However, no discussion about water efficiency would be complete without some consideration of 'virtual water' and the 'water footprint'. These terms are used to describe the water that is consumed in the production of goods and services. The amount of water used for these purposes vastly exceeds that used directly by you in your home.

For example, 140 litres of water are required to grow enough coffee for a single mug full, and 16,000 litres of water are required to produce 1kg of beef. Whilst these numbers seem worryingly big, in some instances, this use of water can be fairly sustainable if food is being produced in areas where agricultural water use is efficient and there is plenty available. Indeed, the majority (70%) of agriculture worldwide is rain fed rather than relying on abstraction from surface water or groundwater, often relatively benign in environmental terms. Getting 'more crop per drop' will be increasingly important in the future in order to prevent water and food scarcity.

Additionally, it's not just food consumption, water is vital to the production of most goods and services. In the UK, our average individual water footprint (the total amount of freshwater used to provide goods and services for a person) is 1245m^3/year. Compare this to the average domestic water use (around 60m^3/year). A region is said to be under 'water stress' if it has less than 1700m^3/person/year available.

If you are interested in wider ecological issues rather than simple domestic water efficiency, you should look at other areas of your life and how your use of goods and services contributes to your water use. Information sources on virtual water and the water footprint are given in the Further reading section.

Summary

Water efficiency should be the first measure considered before resorting to alternative supplies. Similarly, behavioural changes are preferable to changes in appliances and can significantly decrease water use from the UK average of around 150 litres per person per day. As the largest consumer of water in the home, the toilet is one of the first considerations when trying to upgrade or modify water-using appliances. In overall environmental terms, focussing on reducing hot water use is more important than cold water. Water efficient alternatives to existing water-using appliances are available, and may result in other performance benefits.

Further reading

Balancing water for humans and nature. The new approach in ecohydrology. Falkenmark & Rockstrom, Earthscan. An excellent book emphasising the importance of considering 'green' and 'blue' water sources when considering how water scarcity can be avoided.

Building Regulations (2009). Approved document G – Sanitation, hot water safety and water efficiency.

Quantifying the energy and carbon effects of water saving – summary report. Available from www.energysavingtrust.org.uk

Watermark. Final project report to HM Treasury, OGC Buying Solutions. Details findings from a study on water use in non-domestic buildings.

Water resources across Europe – confronting water scarcity and drought, European Environment Agency, 02/2009.

Water UK sustainability indicators. Available at www.water.org.uk
Updated annually.

Websites

AECB water standard and technical guidance, www.aecb.net.

Environment Agency web site (www.environment-agency.gov.uk) has lots
of useful resources including:

   *Water for people and the environment – water resources strategy for England
and Wales.*

   Greenhouse gas emissions of water supply and demand management options,
Science Report SC0700010.

Water efficient buildings website: www.water-efficient-buildings.org.uk

Water efficient product labelling scheme (WEPLS).
http://www.europeanwaterlabel.eu/

Waterwise website (www.waterwise.org.uk) has lots of useful resources
including newsletters, publications, technical notes and research reports
covering most aspects of water efficiency.

Which? The Consumer's Association regularly reports on the performance
of household white goods, including their water efficiency. www.which.net

www.waterfootprint.org – information resource on virtual water and the
water footprint.

Chapter Two
The mains and the alternatives

Sustainability of mains water and drainage – a closer look at the hydrological cycle – assessing a site for a private water supply – quality and quantity of sources – springs, boreholes and surface water – how to get the water out – reinstating old supplies – protecting your water

Before we consider private water supplies, we need to consider the sustainability of the existing mains water system, so that we have a benchmark to compare the alternatives to. A related issue is the sustainability of sewage and rainwater disposal systems; if you are considering rainwater or grey water recycling for water supply, you will be decreasing your impact on drainage systems.

The environmental, economic and social impacts of mains water supplies and drainage systems vary from region to region, so you should consider your local circumstances and your likely private water source if you are interested in which is the most sustainable option. There is a common, often wrong, assumption that small scale systems that 'close the loop' are inherently environmentally friendly. You may have reasons for wanting your own supply, but you should recognise that if mains water is available then installing a separate system is rarely good from an environmental perspective.

Water supplies and sustainability
Environmental considerations – supply

Supplying water is an infrastructure intensive business, and much of the environmental cost is associated with the installation of the infrastructure rather than with the abstraction, cleaning and supply of each litre of water. Installing an alternative water supply, if mains water is available to you, is therefore an extra infrastructure cost with an extra environmental impact.

There are also simple economies of scale; whilst large scale 'industrial' processes may look very intensive and polluting, they are often much less so than carrying out the same function with multiple smaller systems. Clearly, there will be cases where the environmental costs of mains water supplies are high:

- Where the existing mains network is a considerable distance away from where it is needed.
- Where the water company is over abstracting from the source.
- Where the water company is taking water from one catchment and then discharging it as treated sewage in a different catchment (thereby depriving the original catchment of water).

In these situations a private water supply could be the most environmentally sustainable option. In water scarce areas you should pay particular regard to water efficiency measures; installing an alternative supply may be an option if the source you abstract from is less overstretched than that used by the water company. However, since the water sources in a catchment area are linked, this is often not a simple thing to establish. The local CAMS (catchment abstraction management strategy), available from the Environment Agency, will contain lots of information about the sustainability of the local hydrological cycle.

Environmental considerations - disposal

The sewage that you produce leaves your house and flows along sewers to a treatment works. At the sewage works, nutrients and pathogens are removed before the (now clean) water is discharged back into the hydrological cycle. Unfortunately, in many areas of the UK, rainwater enters the same sewers. Sometimes this is due to infiltration from the ground, where sewer pipes have become cracked, or water entering around the joints. Even worse is the large number of deliberate connections of rainwater and surface water drains into the sewerage system. The consequence is that rainwater gets mixed with domestic sewage and industrial waste water and must be treated as sewage at the sewage treatment works.

Additionally, the sewerage network and the sewage works are often not able to deal with the massive increase in flow that occurs during heavy rainfall (the flow within a sewer can increase 10-20 fold during a storm). This can result in untreated sewage being discharged directly into water courses in order to avoid flooding the sewage works; known as combined sewer overflow.

Surface water connections into the sewerage system are no longer permitted, but it will be many years before the mains infrastructure keeps rainwater and sewage separate.

Conventional rain water drainage systems have the additional problem that they collect and channel water from several sources into a single stream, thereby increasing the flow during high rainfall, potentially increasing the risk of flooding. This is completely unlike the natural hydrological cycle, where water would infiltrate into the ground close to where it fell, or flow slowly over a ground surface until it met a watercourse.

Nature's approach is now being mimicked by the installation of 'sustainable drainage systems' (SuDS). These systems take the rainwater from buildings and impermeable surfaces, slow down its flow and filter it before discharging it back into the environment. Where the soil is sufficiently permeable, water is discharged to ground via soakaways, thereby replenishing groundwater supplies and mimicking the natural hydrological cycle.

In areas where the ground is naturally very wet and water won't soak away, SuDS systems that incorporate tanks or ponds are used to slow the flow, before discharging the water at a slower rate into a watercourse. Consideration of your rainwater drainage is therefore also a part of incorporating your building into the hydrological cycle in as sensitive a way as possible.

Economic sustainability

Installing your own water supply will rarely pay back the initial investment or the running costs if you already have mains water. A number of factors are involved:

- Current charges - if your water isn't metered, then decreasing your mains water use by supplementing it from another supply obviously won't save you money. Check what the volume charge is from your water company; it varies enormously across the country.
- Demand for non-potable water. Typically this is higher in buildings used by lots of people during the day (when the water demand is largely

for toilet flushing), or where water is used for other purposes, such as industrial processes. Lower grade water from a private supply can be cheaper than mains water in this instance.

- Is your project a new build or a retrofit to an existing property? It is often more expensive to retrofit water supplies than to incorporate them in the initial design.

- Is there a requirement to limit surface water run off during storm events (often a requirement of planning permission)? If tanks are required to do this, then it may make sense to reuse this water.

- Community scale systems can be more economical, as the infrastructure cost can be divided between more households. However, as you will see in Chapter Three, private water supplies serving more than one dwelling are subject to a much stricter regulatory regime.

Social sustainability

Social aspects of sustainability are notoriously difficult to measure and the following points are far from exhaustive.

Installing private water supplies for demonstration or educational purposes in public buildings may have some merit, since it may result in improved water awareness and encourage wider uptake of water efficiency measures by visitors. However, one could equally well argue that the feeling of water being 'free' once a private water supply is installed might actually increase water use.

Given the complicated nature of the control units and the risk to public health in the event of system failure, the social benefits of private water supply installation are questionable. Indeed, the risk of contracting a disease from a private supply is much higher than from public mains, and most private water supplies are contaminated at some point during the year.

Costs – if a supplementary supply ends up costing the householder more than a conventional mains water supply then they are certainly not appropriate for social housing projects.

The necessity for basic maintenance, such as cleaning filters and checking tanks can be considered as a cost (because of the time taken) or a benefit (in terms of putting people in touch with their resource use) depending on your viewpoint.

Finding a private water supply – assessing the site

Having got a good idea of the availability and sustainability or otherwise of your local mains water supply, you are now in a position to compare it to potential private supplies.

The hydrological cycle and site assessment

The hydrological cycle discussed in the Introduction tells us the most likely sources of water available to us.

Sea water is not an appropriate domestic water supply in the UK. Whilst it is plentiful, purifying it is energy intensive and the resulting water is unpalatable.

Surface water (lakes, streams, rivers, reservoirs and canals) arises either from groundwater (where porous rocks reach the ground surface) or directly from rainfall and runoff. If it arises from rainfall then the amount of water can vary considerably according to recent weather and the size of the land area that provides the catchment. The quality of surface water is highly dependent on the weather; run off from the land will contain potentially harmful micro-organisms and lots of soil particles. In peaty moorland areas it can also be very acidic.

Rainwater as a direct source of water for domestic water supplies is considered in Chapter Six.

Fog and dew harvesting are possible, but are best suited to countries with extreme water scarcity and suitable diurnal weather patterns.

Groundwater is water after it has soaked into the ground. Water can exist in the ground wherever there is space between rock or soil particles. This can either be when the ground is 'unconsolidated' (like soil, gravel, or sand), or when it is a rock with lots of small holes or cracks in it; 'consolidated' (like sandstone, shale or limestone).

Groundwater is generally much cleaner than surface water owing to the filtering that occurs as it passes through rock and soil particles. This is particularly true of deep groundwater (because more filtering has generally occurred) and of groundwater in unconsolidated formations. Rocks that can hold water and allow it to flow through them are known as *aquifers*, rocks that do not are *aquitards* or *aquicludes*. Figure 2.1 shows some of the critical features of groundwater, its movement, and ways to access it.

The movement of groundwater is hard to predict, since water is subject not only to gravity, but also to pressure from surrounding rocks and to capillary action. The boundaries between rocks of different water holding abilities will rarely be straight lines, so the water table (the uppermost level of ground that is water-saturated) is rarely horizontal over significant distances. Unless it enters the sea directly through porous rocks on the seabed, groundwater will eventually surface somewhere on land; sometimes these points are *springs*.

Other groundwater will flow directly into a river whose banks are permeable, so the water level in the river is the same as that in the ground. If you drill into an aquifer you may create a *water table well*, whose level is the same as that of the aquifer it is drilled into. In other cases, the water level in your drilled hole may reach a level higher than the upper boundary of the aquifer, in which case it is known as an *artesian well*. This occurs if the aquifer is confined; an aquiclude above the aquifer prevents the groundwater level rising but if this aquiclude is punctured, this barrier is released. The area that an aquifer refills from is known as the recharge area, and the points/ areas from which it loses water are known as the discharge areas.

The amount of water available from a groundwater supply depends on a number of factors. The first is how extensive the aquifer is. If you find a *perched aquifer* resulting from small deposits of impermeable rocks forming a basin, it may run dry regularly. Even if you bore into a large aquifer, you may be removing water from it faster than it can be naturally recharged.

The recharge of the aquifer may be heavily dependent on seasonal rainfall if it is fairly shallow, but it will also depend on the hydraulic pressure from water in the surrounding rocks, and how well they transmit water, in other words the permeability of the rock. Clay is an extreme example of this; whilst it contains a large amount of water, it is bound tightly to the clay particles so the permeability (and therefore rate of recharge) would be very low.

At the other extreme, geology that includes large fissures (such as limestone) will have a very high permeability. Changes in the amount of water in deep aquifers may be buffered over several years and the water may have entered the ground many miles away. Only if the depth and extent of an aquifer have been completely mapped can you really have any certainty of how much water can be abstracted from it.

Quantity of groundwater will not be your only concern. Its *quality* will vary according to depth and rock formation, as described above. Groundwater will dissolve salts, minerals and gases within the rock. In general, the longer the water has been underground, or the greater the depth it comes from, the more chemical contaminants it is likely to contain, but the fewer micro-organisms.

Independent sewage treatment systems (such as septic tanks and leach

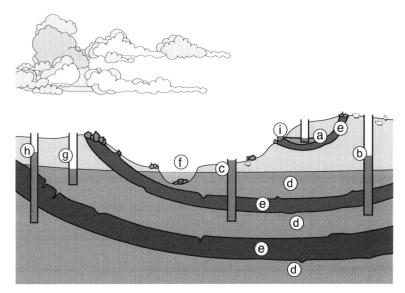

Figure 2.1. Groundwater. The following features are illustrated. (a) perched aquifer;
(b) artesian well (non flowing); (c) artesian well (flowing); (d) aquifer; (e) aquitard; (f) stream
fed by groundwater; (g) water table well (shallow); (h) well/borehole (deep); (i) spring.

fields) are potentially a major source of micro-organisms in the immediate area. Separation distances between sewage treatment systems and groundwater abstraction should be 30 metres as an absolute minimum, and if there are likely to be sewage systems within the general catchment then this should be incorporated into the risk assessment and consideration of treatment technique, as discussed on page 65 in Chapter Three (see table 3.3).

Man-made chemicals such as nitrate fertilisers and organic pesticides will percolate into groundwater and you will need to know where the aquifer is being recharged from and whether or not that area is at risk of pollution. It is also possible to affect the quality of the aquifer by abstracting water from it; the water that moves in to take its place may have different characteristics, or may be saline.

Looking for water

Once you understand the theory behind where water might be on your site, the best way to assess a site for potential water sources is to consult local experts, and then complement the existing knowledge of a site by assessing it from the ground and by looking at maps.

A wealth of hydrogeological data is available on the British Geological

Survey website, ranging from descriptions of the underlying geology, to drilling records from local boreholes. These include the thicknesses and types of all ground that the borehole passes through and provide a much more detailed picture than is available from a two-dimensional map of the dominant geological type.

If you are going to rely on a drilling a well or borehole into a groundwater supply, you will probably decide to get a specialist local water engineer to do the work. They will generally do everything from site assessment to accessing the source and laying the pipe, so you needn't worry about exactly how to find the water.

People to ask

Try and find out where your nearest neighbours get their water from, and whether they have had quantity or quality problems with it. If there are wells nearby find out how deep they are (both resting water level and total depth), how much water they yield and what the water quality is like.

Older people will often remember the sites of wells that have since fallen into disuse. Local farmers will know how soil conditions change over the seasons and what the level of the water table is. They may also be able to tell you if any effluent is being discharged into the ground near your site and the location of any land drains.

Get in contact with the local Environment Agency Water Resources Officer. You will need to talk to them about abstraction once you have found your source, but they may also tell you what the best local sources are, how much water you'd be allowed to abstract, what's likely to be in it and whether or not there is anybody else abstracting from the source.

The Environmental Health team in the local authority may also be helpful; whilst their role is mostly to advise on water quality of existing private supplies, they will be aware of the nature and range of private water supplies in the local area which they are visiting, sampling and risk assessing.

Signs on the ground

Have a look around the site armed with as large a scale map as you can get. You will need to work out where water is, where it comes from and where it goes to, in other words how your site fits into the hydrological cycle. Start by marking out all surface water. Include streams and rivers, pools, and marshy ground.

Walk upstream along any surface water to determine where it starts from;

if it is from a fairly clearly defined spring it may be better to take water from the spring directly rather than from the surface water, as the surface water has had more opportunity to become contaminated. Very often however, the point at which groundwater reaches the surface is a widespread area of boggy ground that isn't easy to capture. Have a look at the types of vegetation on the site as these will often indicate water close to, but not reaching the surface. Plants such as ferns, sedges, rushes and reeds indicate wet ground.

Looking at the site from above can be helpful, either from a nearby hill, or by referring to an aerial map. These may allow you to detect geological features such as fault lines, surface rock features, erosion in places you wouldn't expect it, and tell-tale changes in soil type or the darker lusher green of vegetation that may indicate a spring.

The map will also allow you to work out how big the catchment area is for any surface water on the site. The catchment area provides some in-built storage, which slows water run-off into streams and keeps water flowing when it isn't raining. Bigger catchments will provide more of a buffer than smaller ones, and catchments with steep slopes, exposed rock, shallow soil and little vegetation will have rapid run-off and so provide less of a buffer. Are there any land drains in the area? Fields that are artificially drained must channel the water to somewhere else, and may be a useable source of water, although the quality isn't necessarily good and they can run dry in the summer.

Dowsing

You may be interested in dowsing for water; that is, searching for underground water by watching the movement of sticks, rods, or pendulums held in the hands. Those who believe that dowsing works have often had a direct experience of either meeting someone who dowses successfully, or trying it themselves.

A believer might say that dowsers are supremely sensitive to some property of water, possibly its conductivity or magnetic effects. The sceptic would say that dowsers look for some of the signs for groundwater discussed above and then subconsciously transmit their hunch to their dowsing rods.

The more rigorously conducted studies on dowsing suggest that some individuals have a dowsing success rate far greater than that expected due to chance alone, and very similar to that of any water engineer who has arrived on site with a drilling rig. You may choose to dowse for water, or ask someone to do it for you, but we would caution against relying solely on DIY dowsing before hiring plant machinery to access any water source.

Legal aspects

The aquatic environment in the England and Wales is controlled and monitored by the Environment Agency (Scottish Environmental Protection Agency, SEPA, in Scotland, Environment and Heritage Service, EHS, in Northern Ireland). In general, if you wish to abstract less than 20m^3 (20,000 litres) of water per day from either groundwater or surface water you will NOT require an abstraction licence, but you should check this with the local EA officer. For larger volumes, there is a specific licensing procedure explained on the EA website.

Licence applications can take several months, and if the application is for groundwater, it may require a groundwater investigation consent and pumping test prior to applying for the licence so the process can take longer. This work is generally undertaken on your behalf by the company drilling the borehole.

If you wish to abstract from surface water, there will be a minimum amount of water that you must leave in the surface water at all times, and you will have to demonstrate that your abstraction will not result in any damage to the local ecosystems. Licences will usually specify a maximum volume that can be abstracted and a review date (normally 12 years), after which the licence can potentially be revoked if the sustainability of the local water resource has changed for any reason.

Exclude sources on basis of quality

As soon as you have access to your water supply, you need to make sure that it is of good enough quality to be treatable to the standards you require. Whilst it's technically possible to treat almost any water to drinking standards, that does not mean it is advisable to do so, and in some instances you may be better off finding another source than cleaning one of poor quality. You can often tell quite a lot about the quality of water by looking at it closely.

Is it clear and does it smell? Removing solids is usually the first stage of cleaning water, so the less present initially the better. If you leave the water to settle for a couple of hours you may be able to tell the difference between very small suspended solids and dissolved particles that are colouring the water.

Dissolved substances in the water that cause it to be coloured are not always dangerous, but will at the very least have a significant bearing on how you clean your water; as you will see in Chapter Four. The best treatment for pathogen removal is ultraviolet light, but this is not effective if the water isn't clear. The dissolved substances may be due to organic materials leaching out from peat, or decaying vegetation. This may in turn cause algae to grow in the water, which in turn will support a population of other micro-organisms.

Colour may also be due to the presence of metals, not necessarily in harmful quantities, but they might not be aesthetically acceptable to you. On occasion, crystal clear water can also indicate that levels of some toxic material are so high that nothing can live in the water, but you won't discover this unless you have the water tested. Avoid water that smells; the dissolved gases that result in smells are not always harmful in themselves, but usually indicate the presence of other undesirable substances in the water.

Having looked at the water itself, look at where it comes from, as it may give you a good idea of potential contaminants. This is considered in detail in Chapter Three, in table 3.4. If you have looked at geological maps, you can predict what types of minerals and dissolved salts could be in groundwater, or you may know that there is a risk of heavy metals from certain types of rocks in the area.

Is there a history of metal mining in the area? Might the ground be contaminated by old, capped-off landfill sites? If it's surface water, what are the surfaces it's being collected from? Are they fields where animals graze, or are they planted and farmed intensively? Is anyone else taking water from the source, and, if so, have they had it tested? Are there septic tanks or land drainage systems in the area or any licenced or unlicenced discharges?

On most sites you will only have one potential water source, but if you are lucky enough to have various options, these preliminary indicators of water quality should allow you to establish which source to use. Once you have established this, you should get a sample tested (as discussed in Chapter Three) before investing in the infrastructure to get the water out.

Measuring how much is available

Once you have identified a source of suitable quality, you need to determine whether there is enough of it.

Groundwater: The amount of groundwater available from a source is dependent on the size of the aquifer that you have accessed, and the rate at which it is refilled (known as the recharge). Clearly, if you remove water from an aquifer faster than it can be recharged, your source will dry up. Groundwater investigations are ideally carried out in late summer when the groundwater level is likely to be at its lowest; if there is water present after a long dry spell, you can be reasonably certain that the aquifer will yield water all year. Given that it isn't always an option to wait until summer, boreholes are usually drilled to a conservatively deep level.

Wells and boreholes: A basic *pump test* will determine the likely yield of your aquifer, and is detailed in the box below. These tests are done by specialists, and the local EA officer may require a specific type of pumping test done before allowing you to abstract water. This is only likely to be necessary for large abstractions, or where significant installation costs and risk are involved.

Box 2.1. Pump tests

A pump test is done to establish how fast water flows from the ground into a well or borehole. It requires water to be pumped out in a controlled way, with the resulting effects on water levels being measured. As you remove water, the water level goes down in the immediate area. This effect on water level is known as *draw down*, and the dry(er) area it causes takes the shape of an upside down pyramid, known as a *cone of depression*.

For large abstraction volumes, the impacts of the abstraction on water levels in the surrounding area and on the water environment in general will be measured over a longer period (weeks or months), particularly if there are other abstractions occurring from the same aquifer system. Various types of pump test can be performed depending on what type of data is sought. Samples for chemical analysis may also be taken during the pump test and these can indicate from where the groundwater source is being recharged. A simple pump test for an existing well or borehole serving a single household is just to pump the daily predicted water use out, and check that the recharge rate is less than 24 hours.

Springs: The simplest way of measuring the flow of a spring is using the 'jug and stop-watch' method. Simply hold a measuring jug or bucket under the spring and time how long it takes to fill. You may need to dig around the spring outlet a bit and use some kind of rubber or plastic sheet as a skirt around the area to make it easier to funnel all the water into one place.

Surface water

At a very simple level, you should try and find out if the surface water course has been known to dry up. Even if it appears dry, if you dig away in the stream bed a bit you may well find that it is still flowing underground. You can measure flow in your stream regularly over a period of time to get an idea of how 'flashy' it is (how quickly it responds to rainfall). You can then compare your own data to data from gauged water courses in the area (look at the National River Flow Archive on the internet). The numbers will be different, but the overall pattern of high and low flows in watercourses in an area will be similar.

For larger systems a computer program (LowFlows) is available that incorporates flow data from gauged catchments with estimates for other catchments based on factors such as their total area slope and underlying geology. Very low flow rates can support private supplies; a water supply flowing at just 0.5 litres/minute equates to 720 litres a day.

Getting the water out

Once you have ascertained that there's enough water and it's of a suitable quality, you need to determine how best to get it from your source into a pipe and from there to your house. It's important to keep the water as clean as possible during this process, not just so that you don't have to clean your water more than absolutely necessary, but also because you don't want to contaminate the source; particularly with groundwater. The measures taken to prevent contamination occurring are known as *source protection*, and their importance is emphasised in the Private Water Supply Regulations as discussed in Chapter Three.

Springs

Spring water is an excellent source if it is from deep groundwater; it will be cleaner than surface water, and, since it reaches the ground surface naturally, you may not need to pump it. However, very shallow springs can be much more prone to contamination, as there will have been little filtering of water through the ground. If you are accessing a spring, you will need to build a structure around it to protect it from surface water contamination. This is known as a 'spring box', and is illustrated in figures 2.2 and 2.3.

The spring water enters this box via deliberate gaps in the block work, through a gravel base, or via a pipe buried in loose material behind the box. The box allows some storage to buffer variations in demand and protects the spring from contamination. An area extending at least 4 metres around the spring should be fenced off to prevent access by animals. You should dig an interceptor ditch around the uphill side, to channel surface water away from the spring.

The spring box can be built out of block work, concrete rings or plastic tubes, should extend above ground level, and must have a secured lid, a

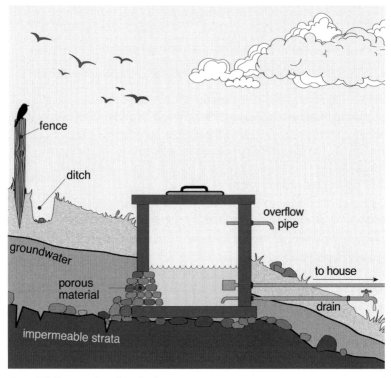

Figure 2.2. Cross-section of a spring box.

vermin proof overflow and a drain as well as the water outlet. The critical feature during installation is usually keeping the area dry enough to work in, and anchoring the structure into the surrounding ground sufficiently to be sure that it won't float away. The overflow will flow more or less continuously, and should be generously sized (so that the cover of the spring box doesn't float off if the flow rate from the spring goes up).

You should not use any pressure within the groundwater to pump the water to height; it is important not to change the natural hydraulic pressure gradients within the groundwater, which could potentially cause the eye of the spring to move. It would also provide a contamination route into the groundwater. The size of spring box depends on other storage components of your water supply system as discussed in Chapter Five.

If the area surrounding the spring is used for animal grazing, your fencing may be preventing the animals accessing their usual drinking water source (and most fences are no match for 600kg of obstinate cow). It is therefore worth considering providing a drinking water trough (for example from the overflow pipe) at a suitable distance away from the spring. This considerably decreases the likelihood of animals attempting to enter the fenced off area and hence the amount of faeces entering the supply.

Box 2.2. Case study – an isolated cottage in Wales served by a spring

Figure 2.3 shows a spring box serving a 2-bedroom house on a remote Welsh hillside. The nearest water main runs alongside a road a mile and a half away. The cottage is on a steeply sloping site that forms a slight valley in a large grassy field. A conspicuous boggy patch in the field indicated the presence of a spring, fed from the large catchment above. This area been developed into a spring box which serves the house. It is built of bricks on a slate slab base; the outlet pipe is 25mm in diameter and is covered by a coarse strainer. The volume of the spring box is about 1.75m³, and it is situated about 10 metres above the house. There is therefore sufficient water pressure to feed the house without needing a pump. The spring box therefore functions as both a store of water to cope with peak demands, and as a header tank. The residents are water efficient and the spring has never run dry.

Figure 2.3. Spring box, Wales. This spring box serves a 2-bedroom house. There are spaces between the blocks underground on the upslope side, allowing water to enter the spring box. The water pipe leading to the house leaves the spring box at ground level at the front of the structure; the upper pipe is the overflow. The spring box would be improved by a more tightly fitting lid, an interceptor ditch and a fence to prevent access by animals. The overflow pipe overflows to ground, causing the boggy area in front of the spring box.

Modified spring designs

At some sites there will be a distinctive 'spring line' where water emerges over a more dispersed area at the same height across a hillside. In this instance, a modified structure of a spring box collecting from an infiltration area (figure 2.4) will maximise the water yield.

At its simplest, a spring can be captured and safeguarded simply by driving a pipe horizontally into the eye of the spring, and fencing around the area to prevent contamination from surface water and animal faeces. It is important that the pipe itself does not allow contamination into the water source (so it must be sloping downwards, and should preferably have a trap). This solution is sometimes known as a 'horizontal well' and is commonly

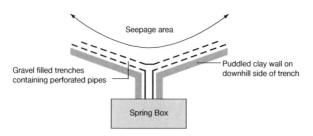

Figure 2.4. A spring with infiltration area (plan view). This technique maximises yield in situations where the spring hasn't got a single 'eye'.

used by farmers to convert boggy patches in fields into useful water supplies for animal drinking troughs. It is somewhat trial and error, as the pipe can get blocked with silt fairly easily and it can be difficult to know where the underground pipe is in relation to groundwater.

Wells and boreholes

If your groundwater supply isn't bubbling to the surface in a spring, you will need to sink a well or a borehole. Wells and boreholes have similar features, but the dimensions are different. A well is generally of a large diameter (at least a metre), dug by hand or mechanical excavator and lined with concrete rings or masonry. Boreholes are smaller in diameter (6 inches is usual for domestic) and are drilled, and then lined with steel or plastic tubes. If you are accessing water from below 4-5 metres, a borehole is likely to be more economic and easier to build than a well.

Both wells and boreholes are classified as 'shallow' or 'deep'. This is not a reference to how far below ground level they extend, but rather to what level of aquifer they are abstracting water from and both are illustrated in figure 2.1. Shallow wells and boreholes are more likely to dry up (or at least suffer from large changes in water level) than deep ones. Since the water from deep boreholes has potentially come from a catchment several miles away, it can be difficult to predict what contaminants might affect it.

The major features of constructed wells and boreholes are illustrated in figure 2.5. The point at which groundwater reaches the surface is particularly prone to contamination and this must be prevented, either by having the top of the borehole projecting above the ground surface, or the area around the top sloping away from the borehole (both illustrated). Where neither of these are possible, the head works of the borehole are contained in a waterproof manhole that prevents ingress of surface water. The borehole should not

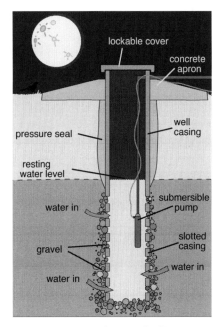

Figure 2.5. Important features of wells and boreholes.

be sited in areas where serious contamination could result from accidental spillage, such as in car parking areas, or near heating oil tanks.

Contamination may also come from water in the topsoil and subsoil, so the solid lining should extend down the sides of the well/borehole into the aquifer. Below this point perforated lining is used. The central shaft provides an amount of water storage that can act as a buffer between the demand (water pumped out for use) and the supply (water seeping in to the borehole). There may well be gravel both at the base of the well/borehole, and around the perforated pipe, to prevent silt particles entering the shaft. Note that since figure 2.5 illustrates the features of both wells and boreholes, the dimensions are a bit misleading; in a borehole the pump fits quite snugly (often only 50mm clearance) into the shaft.

Most people will leave well digging and borehole drilling to a professional contractor (as boreholes are often 60 metres deep specialist machinery is required) and you should refer to the Environment Agency guide to ensure that all works are carried out according to best practice. Many professionals are members of the Well Drillers Association.

If you are interested in digging a well yourself, detailed information is available in *Hand dug wells and their construction*, details of which are listed in the section on Further Reading. It is possible in some instances to combine a borehole for private water supply with a 'standing column' ground source heat pump as part of the same installation.

Reinstating old groundwater supplies

If you have an abandoned water supply that you would like to reinstate, the first thing you should do is try and establish why the source was abandoned in the first place, to see if it's worth putting any effort into. Did it run dry, was it of poor quality, or did it simply fall into disrepair when mains water became available? Depending on what you find from these initial enquiries you may choose to unblock the well.

How easy this will be depends upon how it was sealed off in the first place. It may have been simply covered with a slab and then become overgrown, or it may have been filled with rubble, concrete or even rubbish. If the well has been filled in, it may be easier to dig a new one into the same aquifer than to reinstate the old well. Never go down a well to investigate its condition without taking adequate health and safety precautions; collapse of the walls, unknown depth, unclean water and stagnant air are all possible.

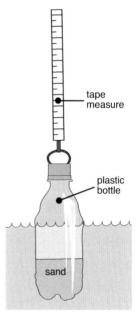

Figure 2.6. Home made dipper. You can test the water level in an existing well using a 'dipper' made out of an old plastic bottle with sand in the bottom, attached to a tape measure. The sand weights the bottle slightly so that when you lower it into a well, you can feel the change when the bottle hits the water.

Once you have accessed the old supply, you should find out how deep it is and what the water level is. You can check the water level by lowering a 'dipper' into the well. The professional versions consist of two electrodes on the end of a long tape measure; you lower the dipper down the well and when the electrodes hit the water surface a circuit is completed that makes a buzzer go off. You can improvise a dipper using a 250ml bottle with some sand in it (so it floats upright), tied to a measuring tape or piece of string (figure 2.6).

You will be able to 'feel' the water surface by the change in weight on the end of the line. Take a measurement at this point. Next, take something that will sink and attach it to the measuring tape, to determine what the bottom level of the well is. You can then calculate the amount of water in the well.

Pump out any water that is in the well; if it hasn't been used for years then the water it contains will not be representative of the water that will recharge it. The next thing to do is to see if the well refills once it has been pumped out. If it doesn't you should examine the lining of the well at the bottom; it may have collapsed, or become blocked up with silt so water can't enter it any more.

There may be chemical precipitation or bacterial slime on the lining that has decreased its permeability. These problems can be cured (by mechanical scrubbing or by acid cleaning), but these are jobs you will usually leave to specialists. If having checked these possibilities, the well still doesn't refill, then it's probably not worth going to any more trouble and you should look at digging a new well or borehole.

Assuming the bottom of the well has refilled, you should get the water analysed to see what's in it. Even if you want it for garden watering rather than for drinking, you will want to check that it hasn't got high levels of metals or salts in it. Once you have determined that the *quality* is suitable, you will need to ascertain the available *quantity*. This will require a pumping test, as described in the box on page 44. Having been through the process of discovering what the water quality is, and ascertaining that there is enough of it, you should undertake a thorough examination of the lining and superstructure of the well, and either upgrade or rebuild, incorporating the features illustrated in figure 2.5.

Streams or rivers

How you choose to take the water out of a stream or river depends primarily on how much the water level changes and what the water quality is like. If you just put the end of your pipe in a stream and weigh it down, the pipe can get blocked fairly quickly.

This can be very troublesome; the blockage in the pipe could be anywhere, and since the rest of the pipe will probably be buried underground, the best option for trying to remove it is to force water at pressure back through the pipe. It is far better to stop pipes blocking in the first place by using an inlet strainer.

DIY versions of these are a false economy and a stainless steel mesh (like a tea strainer) should be used in preference (figure 2.7). These can themselves get blocked, but at least this will be in a known and easily accessible location. This inlet strainer is vulnerable to damage, so should be protected. There are three main methods, of varying degrees of sophistication.

A. Rock filled gabion in the stream (figure 2.8). The end of the pipe with the strainer on it is contained within a rock filled basket to both prevent it moving and protect the strainer from damage. Try and position the inlet in a straight part of the stream to minimise the amount of solids that will get

Figure 2.7. Good (left) and not so good inlet strainers for water pipes.

swept in by turbulence.

B. Inlets built into the bank. Excavating part of the bank (as shown in figure 2.9) so that the inlet pipe faces downstream will reduce the amount of solids entering the pipe. You should reinforce the bank around the inlet with gabions or large rocks.

C. Infiltration galleries (figure 2.10). This is a more sophisticated strainer, sometimes used when the river contains a lot of silt. An upturned tank (for example a domestic header tank) containing gravel can be buried next to the stream, or under the stream-bed if it is loose; you can build a temporary dam to keep the area dry whilst you are working.

A variation on an infiltration gallery is to dig a well adjacent to the surface water supply and use the natural filtering properties of the river-bank to remove much of the suspended solids that are contained in surface water sources. The void of the well also provides some storage capacity and should have the features illustrated in figure 2.5.

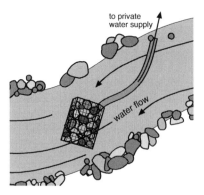

Figure 2.8. An inlet pipe with a rock-filled basket surrounding it to provide some coarse filtration.

Existing lakes and reservoirs

Existing lakes and reservoirs are straightforward sources of water to access for private supply. You will need to monitor the water level over time to see how much it changes. You should find out how and where the lake fills and empties from; in some instances you could be better off collecting from one of

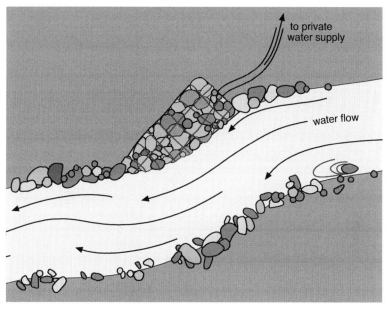

Figure 2.9. Inlet built into the bank of a stream. A stone-filled wire basket (known as a gabion) is buried in the bank, with the end of the inlet pipe inside it. This acts as a filter to prevent debris entering your water supply.

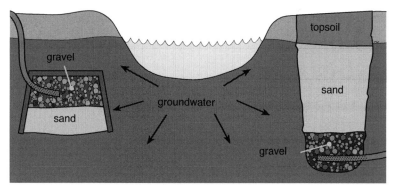

Figure 2.10. Infiltration galleries. When a stream or river is a result of the water table reaching the surface, the banks of the watercourse will be permeable, allowing water to be naturally filtered through the bank before it enters your water supply. The system on the left consists of an upside down tank, which can either be buried in the bottom of the stream, or to one side of it. In both cases you will need to divert the water during construction.

these sources rather than from the actual lake. Is there much surface water run off into the lake that will affect water quality? Lakes which suffer from algae should be avoided. Outlet structures should take advantage of natural settlement and can consist of an outlet pipe with a strainer held above the lake sediment by a float and anchor (made on a similar principle to that shown in figure 6.7 on page 146, but with the addition of an anchor and a more robust float).

Box 2.3. Case study – the Centre for Alternative Technology

CAT is situated on the site of an abandoned slate quarry in mid-Wales. In the middle of the 1800's, the slate miners built a reservoir, dammed with slate and soil, so they could drive a water wheel, which in turn powered slate cutting machinery. The reservoir is fed by two streams, which are fed from rainfall on unimproved upland grazing land, and has a capacity of around 5 million litres.

Water exits the reservoir via a 200mm diameter pipe, which runs across the hillside for 200 metres before the drinking water is 'T'd off in a 63mm pipe. This water passes through a slow sand filter (discussed in Chapter Four) and into storage tanks (with a total volume of around 7m³) that provide a buffer to cope with peak water demands. It then flows down a 63mm pipe to the level of the site (some 30 metres below), where it passes through a manifold containing 4 ultraviolet bulbs that kill any remaining pathogens in the water (UV for water treatment is discussed in Chapter Four).

The remaining water, not used for domestic purposes, continues along the hillside in the 200mm pipe before dropping down to the level of the main site through a hydroelectric turbine and into another reservoir. From this point the water either flows down through a second hydroelectric turbine 30 metres below, or is used to drive a water-balanced cliff railway. When a new residential building was built on the main site, insufficient water pressure and storage capacity in this system led to a parallel system being installed. For this part of the site, water from the reservoir passes through a bag filter (discussed on page 82) to remove solids, and then through a UV to kill pathogens. The reservoir itself is effectively the storage tank.

If there isn't enough in the stream – building a reservoir

If the flow in a surface water source is too variable to be a reliable supply, you may be considering building a reservoir. Since this can seriously disrupt the aquatic life in the water source, the Environment Agency must be consulted and they will probably limit the amount of water you are allowed to remove. This will usually be based on maintaining dry weather flow in the watercourse, so you must ensure that the excess that you can abstract is sufficient for your needs.

Building reservoirs is best left to a professional. They will calculate the necessary size, carry out the excavations and install the drainage system for you. Reservoirs are expensive to build, particularly if the local soil is not sufficiently impermeable to provide a lining and build the dam with. There are two basic types; those that involve damming a watercourse (impounding reservoirs) and those filled by the stream but dug adjacent to it (off-stream reservoirs).

Source protection

When dealing with private water supplies, you should do all you can to protect your water source. Specific source protection mechanisms for the various structures you may build around your water source have been described earlier in the chapter, but the following general points should be borne in mind.

Actively minimise water use: depending on the nature of the source, there may be a risk of it running dry, this will be obvious with surface water, but you may have little warning that this is going to occur with groundwater sources. This is not just a problem of insufficient *quantity* available to you. In the process of running dry, the water *quality* may decrease considerably (for example as you start using more stagnant water from the bottom of a lake), which may mean it needs more treatment. Running it dry will also seriously damage the ecosystem from which you are taking the water; if there is a risk of this occurring the authorities may put limits on the amount of water you are allowed to abstract.

Even if it appears plentiful, it is good practice to be as economical as possible with your water, as described in Chapter One. You should also be aware of who else is abstracting water from the same source as you, and what the consequences of level changes are to them; depending on the local geology, they may run out of water before you do.

Not making what you are about to use dirty: If you are going to a lot of trouble to find your water source, it clearly makes sense to try and keep it clean whilst you are abstracting it and moving it to where you want it.

Not contaminating your source: Whatever structures you put in or near the water source to enable you to collect it, you must make sure that they are clean, and are certified for use with potable water sources. Source contamination is a particular concern with groundwater supplies. If you are opening up a groundwater supply to the surface that was previously sealed, you could potentially pollute an entire aquifer, for example with the pathogenic organisms found in animal faeces that may be present in surface water run off, or from chemicals leaching out of your pipe or water tank. You must also prevent backflow of water into the groundwater source (since it may have been contaminated whilst above ground level); for the same reason you cannot put water back down boreholes or into groundwater without licencing and adequate treatment.

Summary

Your main concerns when choosing a private water supply will be the quantity and quality of the water available to you. Your site may have a number of sources available, in which case the following is offered as guidance on which sources are most likely to be appropriate:

1) Mains supply – clean, plentiful and cheap!
2) Spring – moderate costs to access, may run dry in summer.
3) Well or borehole – expensive to access, but good and consistent quality and quantity.
4) Surface water – easy to access, but worse quality than groundwater.
5) Direct rain – considered in Chapter Six.
6) Reused water – considered in Chapter Seven.

Further reading

British Regional Geology series. Books covering the whole of the UK. Published by the British Geological Survey.

Oxfam Water Supply Scheme for Emergencies.
Instruction manual for Hand Dug Well Equipment.
Available to download from the Oxfam website (www.oxfam.org.uk) – designed for field engineers using Oxfam emergency kits, but provides useful detail on wells and boreholes.

Field Hydrogeology, Brassington. Wiley, 1998.
Details on site assessment, pumping tests for estimating yields of boreholes.

Finding Water, Brassington. Wiley, 1995 (second edition).
Out of print and out of date in places, but still worth reading.

Hand dug wells and their construction, Watt & Wood, (ITDG), 1979.
Useful guide if you intend to dig a well yourself. Available from CAT Eco-store.

Water supply borehole construction and headworks.
Guide to good practice. Available from EA in paper copy, or downloadable from their website.

Websites

Environment Agency Water Resources Officers. Local officers deal with licensing private supplies and are a valuable source of information on local hydrology. They produce a number of leaflets on good practice when constructing wells and boreholes. Local contact details available from www.environment-agency.gov.uk

Meteorological Office. Has data on rainfall and other weather conditions across the UK. www.met-office.gov.uk

Centre for Ecology and Hydrology. The main research organisation in hydrology. www.ceh.ac.uk

Catchment Abstraction Management Strategies (CAMS). Detailed local level assessments of the sustainability of the water environment, available from the Environment Agency. www.environment-agency.gov.uk

British Geological survey. Records of boreholes in England and Wales. http://www.bgs.ac.uk/data/boreholescans/home.html

Chapter Three

Water treatment – underlying theory and regulation

Deciding on water quality requirements – regulations – risk assessment – water sampling and testing – microbiology – checks to do on your water supply

Once you have found your water, you will need to decide how best to clean it. You are around 50 times more likely to contract a disease from a private water supply than from a public one, so you will need to ensure that you have the right type of system and that it is properly maintained. This chapter deals with the underlying theory of water treatment techniques and the regulatory regime, whilst Chapter Four discusses the practicalities.

There are lots of methods you can use to clean water, and in order to choose the right one you will need to consider a number of issues:

What is the quality of the source water? The treatment technique will depend on what is in the water that needs removing. Water quality parameters can also be related to each other, for example mildly acidic water isn't harmful in itself, but is much more likely to have metals dissolved in it than water that has a neutral pH. For parameters that are toxic by the same mechanism of action in the body, there may be requirements for combinations of parameters in addition to single values; for example pesticides, where there are limits on total concentrations of a range of chemically similar pesticides, rather than separately for individual pesticides.

What water quality do you require? The basic assumption in this chapter is that you will want to treat at least some of your water to UK drinking water standards, and this quality is defined in law under the Private Water Supply Regulations.

Do you require all your water to be drinking water quality? Sometimes it makes sense to have two water supplies of different quality. This might be the case if your source water needs a lot of cleaning, or you use large quantities of water for purposes that don't require high quality (for example garden watering, toilet flushing), or at a different location, or with very different pressure requirements. In extreme water scarce environments, dual water supplies are widespread; much of Hong Kong is served by a dual water supply with one system carrying potable quality water, the other carrying sea water, which is used for toilet flushing.

At what point between the water source and its end use are you going to clean it? The nature of the substances that need removing and the type of source will determine whether the treatment system is installed before or after storage, or (as is often the case) a combination of both. This is discussed on page 100.

How much maintenance does the system require? All systems will need periodic maintenance and/or replacement parts, but the frequency and nature of this varies between systems (and is discussed on page 68).

How can you do all of this in a sustainable way? Your primary concern will be to have a suitable quantity and quality of water, but the environmental impact of treatment processes varies considerably and you may wish to bear this in mind when choosing a treatment technique.

How many people would be affected by a failure in your water supply? For small scale domestic supplies serving single households, simple water treatment systems with only one or two elements coupled with good source control are suitable provided that they are maintained and operated correctly. However, if there are significant numbers of people using the water (for example several properties or a commercial enterprise), regulatory requirements are stricter, and this is reflected in the risk assessment that applies to the supply.

What do we want to achieve by cleaning the water? What quality of water do you require?

The basic qualities of water that we are interested in are health, aesthetics (taste and colour) and corrosiveness.

Health: Water used for drinking and cooking must be safe to drink. Immediate effects of unsafe water are often an upset stomach, but a lack of obvious symptoms does not guarantee safety. You could be slowly poisoning yourself with excessive metals or pesticides, and continued low level infection with certain bacteria can also cause long term health effects including cancer and heart disease.

It's not just drinking water that needs to be clean enough to drink; you may breathe in small quantities of water as a vapour when you wash, and your lungs can be a more vulnerable barrier than your gut lining – as evidenced by the risk of contracting Legionella, for example from air conditioning systems or cooling towers where vapour is produced. The quality of water for personal washing and bathing is particularly important in households with children, who almost inevitably drink bath water occasionally.

Aesthetics: Taste and colour are very subjective variables. You will probably expect the water that you drink to be completely clear, but you may be happy enough to flush your toilet with water that is slightly cloudy. In the same way we have different attitudes and sensitivities to subtleties of taste, be they good or bad. Your attitude to both colour and taste may also vary according to whether or not you know what is causing it; if you knew that your water had a brown tinge because it originated from peaty soil you might look at it more favourably than water that was cloudy because it came from a grey water recycling system. It may also vary according to whether the system is single household, or is public/communal.

Corrosiveness: It is important that your water doesn't cause any physical damage to the pipes that it passes through or anything that those pipes are plumbed in to, such as heating systems and appliances. This damage could be caused by the pH or the hardness of water, by blocking filters in appliances, or by abrasive action of particles in the water.

Different qualities for different purposes

Regulations require water for human consumption to be 'wholesome'. This is defined in chemical and microbiological detail, but basically requires that the water doesn't contain any substance at a concentration that is harmful to health, or any substance that may interact with another to cause it to be harmful. In most instances water for washing and bathing should be considered as included within these requirements for wholesomeness, as well as water for drinking, cooking and food production.

Water used for other purposes, such as toilet flushing, washing machines, heating systems and outside taps is not subject to the same legislation, and need not necessarily be treated to drinking water standard, but it should be clearly labelled to ensure that it isn't mistaken for a potable source. Guidelines from a variety of agencies have been drawn up recommending water qualities for various uses, as shown in table 3.1, although in most instances the emphasis is on risk assessment as opposed to specific water quality values.

Use	Regulation/guidance	Standard
Drinking, washing	Private Water Supply Regulations (2006, 2009, 2010)	Various microbiological and chemical parameters (table 3.5)
Toilet flushing	BS8515 (guidance only)	< 250 E coli/100ml, <100 enterococci/100ml, <1000 total coliforms/100ml
Swimming pool	WHO guidelines, British Standard CoP	<1 coliform/100ml
Garden watering	BS8515, WHO	As for toilet flushing (BS8515) <1000 coliforms/100ml (WHO).
Pressure washers and sprinklers	BS8515 (guidance only)	<1 E coli/100ml, <1 enterococci/100ml, <100 legionella/ litre, <10 total coliforms/100ml

Table 3.1. Water quality for various purposes.

Regulations on private water supplies in the UK

The requirements for private water supplies in Scotland, Northern Ireland, England and Wales are set out in the Private Water Supplies Regulations (which came into force in 2006, 2009, 2009 and 2010 respectively). They are broadly similar to each other, and are an implementation of the European Drinking Water Directive (98/83/EC), which in turn is based on guidelines from the World Health Organisation (WHO).

The international nature of the WHO guidelines means that many parameters that are listed are unlikely ever to cause problems in UK water supplies, so the hierarchy of regulations and guidance allows countries to slim down the regulations to suit national requirements. The PWS Regulations are implemented by the Environmental Health team in the local authority (and by the Department of Environment in Northern Ireland).

The underlying philosophy of the regulations are to combine risk assessment with water quality tests, and to have an overall focus on private water supplies that pose risks to large numbers of people. These supplies are sampled for a wider range of parameters and at a greater frequency than smaller supplies.

Risk assessment

Historically, regulations for drinking water were based entirely on taking water samples and testing them for contamination. However, a water sample is simply a snapshot of water quality at that time; it provides no indication of past or future water quality because contamination of a water supply is usually sporadic, for example only occurring after heavy rain, or depending on the presence of infected farm animals in neighbouring fields.

Indeed a study in Scotland found that the majority of private water supplies contained microbiological contamination at least once a year, but many of these supplies were clear most of the time. Since many private water supplies are only sampled every 5 years or so (if at all), most PWS owners are at risk from their water supplies and may have a false sense of security by regarding a test result as an "all clear".

Consequently, risk assessment is becoming a major feature of the regulatory regime for private water supplies, and a good assessment provides practical guidance on maintenance of a system and the likely consequences of maintenance failures. It also tends to lead to a 'multiple barrier' approach; the presence of a number of safeguards (including source control measures, site issues, treatment systems and maintenance) that considerably reduces the risk if any single element of the regime fails.

The PWS Regulations require supplies to be risk assessed at least every 5 years. However, England, Wales and Northern Ireland have exempted single households where water is not used for commercial or public purposes from the requirements. Effectively therefore, the majority of private water supplies fall outside the regulatory regime. For details on supplies that fall within the regulations, see table 3.2.

If your supply is to a single dwelling and you choose to ask the local authority to carry out water sampling and risk assessment, there will normally be a charge for it. You should also note that if your supply fails the standards set out in the Regulations, they may then take action against you (although whether they choose to do this or not will vary). For this reason, local authorities will make it clear that if you are outside the regulatory system you can get samples tested privately, and may offer informal advice.

Given that most readers of this book are expected not to be subject to formal regulation of their water supply, the following constitutes a personal opinion on the most appropriate approach to water testing and risk assessment for these supplies, and should not be regarded as definitive, and it will not necessarily be the same approach as is taken by a regulatory authority.

Whilst risk assessment and health and safety legislation often come in for a lot of criticism, the underlying principles as applied to private water

Country	Key regulatory specifics
England	**Large supplies:** Supplies on average at least 10m³ of water per day OR serves 50 or more persons OR is supplied or used as part of a commercial or public activity. **Other supplies:** Any supply that does not fall within the above definition, other than private supplies to single dwellings not used for a commercial activity. The local authority must carry out risk assessment of a supply to a single private dwelling if requested to. Further distribution of mains water supplies (in other words sites that are served by mains water, where the water is distributed through privately owned pipes to a number of properties, such as estates, campsites and so on) also fall within the Regulations. These are known as private distribution systems.
Wales	As England.
Northern Ireland	As England, except the local authority is not required to carry out a risk assessment if the householder requests it, but must instead 'offer appropriate advice'.
Scotland	See page 197. All private supplies in Scotland, including supplies to single dwellings, are subject to the Regulations.

Table 3.2. Supplies falling within the Private Water Supply Regulations. If your supply does fall into a category that is regulated, Appendix 1 summarises the Regulations.

supplies make a good deal of sense. The official guidelines on risk assessment for private water supplies are several hundred pages long, but are distilled into a checklist in table 3.3. This comprises four main elements:

• Basic details of the supply.
• Details relating to the site and the likely contamination.
• Source protection details (discussed in detail in Chapter Two).
• Treatment details (discussed in detail in Chapter Four).

The emphasis is on source control to avoid microbiological contamination of water supplies, with treatment as an additional (and necessary) line of defence, rather than the only one. Risks from microbiological contaminants are considered to be greater than those from other contaminants. For dissolved parameters, prevention of contamination using source control measures is generally more difficult, so sampling to ascertain which parameters are of concern, followed by specifically targeted treatment methods is the usual approach.

General record keeping	Comment
Supply category under the Regulations	See page 194.
Contact details for people using the supply	
Contact details for any other parties (for example landowner if source does not belong to householder)	
Contact details of people supplied, and nature of their use	Particularly important for supplies serving more than one property, and where visitors/public are supplied, so that they can be contacted in the event of any supply failure.
Schematic of supply	To include source, properties supplied, intermediate tanks and treatment stages, stop valves, pipe locations if known.
Description of the source, photos where relevant, grid reference	Including nature and condition of infrastructure, cross referenced to the schematic.
Estimated daily volume supplied	Influences regulatory category and sampling frequency.
Details of any treatment systems	Location within system, condition, supplier, maintenance requirements (and see below).
Any test results applying to the supply	Including those done privately, as well as by the local authority.
Dates and details of any investigations	Useful historical context.
Details of any maintenance undertaken	For example cleaning or changing components.

Table 3.3a. Starting point for householders interested in improving their private water supply. Basic details of the supply and record keeping.

Site issues	These influence the nature of the likely contaminants and affect initial choice of water source (see page 37). With existing systems it influences the water quality parameters that might need to be tested for. See page 69 and table 3.4.
Stagnant water or poor drainage around source	This water will be of a lower quality than the source itself and cross contamination might occur.
Livestock access, wild animals	Risk of faecal matter from either could contaminate supply.
Vulnerability to agricultural runoff or other farm wastes	Could represent source of faecal contamination (from muck spreading), fertilisers or pesticides.
Forestry activities	Felling, thinning and planting of commercial forestry can all cause major changes in runoff (volumes and nature of contaminants).
Awareness of private supply by those who may affect it	Are people who might contaminate your supply by their activities (for example farmers, foresters upstream of the source) aware of its existence?
Waste disposal in area	May result in specific less common, contaminants and therefore has an influence on the water quality parameters that should be tested for.
Industrial activities, including mining	As above.
Presence of off mains sanitation	Locations of systems, nature of treatment, include systems discharging to ground adjacent to surface water.
Other groundwater access points	These can be sources of contamination of local groundwater if good housekeeping measures aren't in place at these locations.

Table 3.3b. Starting point for householders interested in improving their private water supply – site issues.

Supply specific issues	Source protection measures applicable to specific supply types
Stock proof fence and interceptor ditches around source	For wells/boreholes and springs. See page 45 and figure 2.2.
Concrete apron around well/borehole	See page 48 and figure 2.5.
Cover/access point	Water tight, and preferably lockable.
General condition of supply	Integrity of brickwork, tank materials, overflow pipes and so on.
General condition of pipework	Particularly relevant where pipes run on ground surface, or which are subsurface and of a material that is prone to damage.
Vermin protection of overflows	Present and intact.
Coarse filter on surface water supply inlets	To prevent silt and organic material entering system. See page 52 and figure 2.7.
Protection of other components	All intermediate tanks and components should have similar protection to that described above.
Backflow protection	Particularly from systems including connections to animal drinking troughs, or where supplies of differing qualities are incorporated. See page 149.
Nature of supply within the house	Many properties have header tanks in the loft feeding some appliances. These tanks are particularly prone to vermin and must be covered, and cleaned out annually. If the tanks supply taps from which water might be drunk, the risks of this need to be considered. See page 61.
Supply variability (quantity or quality)	This is an indicator of how closely connected (and therefore vulnerable to) the system is to surface water whose quality can vary significantly depending on weather conditions. See page 37.

Table 3.3c. Starting point for householders interested in improving their private water supply – supply specific issues.

Treatment system specific issues	In line with manufacturers or installers instructions. The following is a good starting point for systems where instructions aren't available.
Slow sand filter	Remove top 3cm of sand every 4 months or when flow rate decreases. Wash sand and return it to the filter.
Ceramic filter	Scrub clean (using aseptic technique) when flow rate decreases. This varies according to the turbidity of the water, but is often required every 3 months. Replace when the cartridge shows visible signs of being eroded by the cleaning process.
Carbon filter	Replace at intervals recommended by manufacturer (often 6 months).
Steel coil filter	Backwash weekly, remove from housing and clean annually.
Polypropylene, polyester, viscose, nylon filters	Replace when flow rate decreases, or at intervals recommended by the manufacturer (often 6 months).
Ultraviolet	Replace bulbs and clean quartz sleeve at intervals recommended by manufacturer (often annual). Check daily for bulb failure, or consider alarm.
Ion exchange resins: softeners	Replenish salt solution at intervals recommended by the manufacturer.
Dolomite filter	Refill with media at intervals recommended by the manufacturer (under-sink versions), or when quantity of dolomite chippings has halved (DIY versions). Alternatively monitor pH and add dolomite accordingly.
Ion exchange resins	Replace at intervals recommended by the manufacturer (often 6 months).
Other comments/summary	Any other relevant details about the nature of the treatment system.

Table 3.3d. Treatment system specific issues.

Water sampling and testing

Since your choice of treatment technique will vary according to what's in the water, the first step will be to get a water sample tested. But how do you know what you should get it tested for? You will need to strike a balance; most tests are specific to an individual parameter, so won't pick up things that aren't directly tested for, but when there are so many potential contaminants, testing for everything would be prohibitively expensive.

Even if your private water supply is not subject to regulation, it is worth discussing the matter with your local Environmental Health team; they will know from other test results in the area what contaminants to expect and can suggest both what tests are most appropriate and will have a list of testing laboratories. The parameters in table 3.4 are suggested as a starting point. Table 3.5 indicates parameters listed in the PWS Regulations; don't be alarmed by this, most of these contaminants are extremely unusual and rarely tested for, even in large scale supplies.

Situation	Suggested actions/parameters
All supplies	• Microbiological parameters: dependent on lab, but usually a subset of coliform bacteria, E. coli and enterococci. • pH, conductivity, turbidity, nitrate. • Use checklist on page 66 to ascertain likelihood of other contaminants. • Seek advice from local Environmental Health practitioners on specific parameters of local concern.
Where low pH expected	If water tastes bitter or causes unacceptable staining, test for Fe and Mn. Seek advice from local Environmental Health practitioners in case other metals are common locally.
Existing supplies where plumbing system may be pre-1970	Lead.
History of industry, waste disposal sites, mining activities	Extra precautions may be needed. Seek advice from local Environmental Health practitioners on specific parameters.
If conductivity reading from a previous test is high	Seek advice from local Environmental Health practitioners.

Table 3.4. Suggested approach for water sampling in private water supplies.

Parameter	Concentration or value
E. coli	*0/100ml*
Enterococci	*0/100ml*
Chlostridum perfringens (including spores)	0/100ml
Coliform bacteria	*0/100ml*
Colony counts	No abnormal change
Acrylamide	0.1 µg/l
Aluminium	200 µg/l
Ammonium	0.5 mg/l
Antimony	5 µg/l
Arsenic	10 µg/l
Benzene	1 µg/l
Benzoapyrene	0.001 µg/l
Boron	1 mg/l
Bromate	10 µg/l
Cadmium	5 µg/l
Chloride	250 mg/l
Chromium	50 µg/l
Colour	20 mg Pt/Co
Conductivity	*2500 µS/cm*
Copper	2 mg/l
Cyanide	50 µg/l
1,2 dichloroethane	3 µg/l
Epichlorohydrin	0.1 µg/l
Fluoride	1.5 mg/l
Iron	200 µg/l
Lead (until 2014)	25 µg/l

Table 3.5. Parameters and their limits in the Private Water Supply Regulations. Which combination of these to test for is discussed on page 69. The most common parameters tested for are highlighted in bold italics.

Parameter	Concentration or value
Lead (from 2014)	10 µg/l
Manganese	50 µg/l
Mercury	1 µg/l
Nickel	20 µg/l
Nitrate	*50 mg/l*
Nitrite	*0.5 mg/l*
Odour	Acceptable and no abnormal change
Aldrin*	0.03 µg/l
Dieldrin*	0.03 µg/l
Heptachlor*	0.03 µg/l
Heptachlor epoxide*	0.03 µg/l
Other pesticides*	0.1 µg/l
Total pesticides (all *)	0.5 µg/l
PAH	0.1 µg/l
pH	*6.5 to 9.5*
Radioactivity	0.1 mSv/year
Selenium	10 µg/l
Sodium	200 mg/l
Sulphate	250 mg/l
Taste	Acceptable and no abnormal change
Tetra and tri chloroethene	10 µg/l
Tetrachloromethane	3 µg/l
Total organic carbon	No abnormal change
Trihalomethanes (total)	100 µg/l
Tritium	100 Bq/l
Turbidity	*1 NTU*
Vinyl chloride	0.5 µg/l

Table 3.5. Continued.

You should use a UKAS (United Kingdom Accreditation Service) certified laboratory, who will test the samples using standard techniques. You will need to take the sample in such a way that it accurately represents the water quality. There are procedures to follow in terms of choosing the most appropriate point to sample from, avoiding contamination during sampling. Details of this are included with the instructions sent to you by the laboratory undertaking the testing. Most laboratories will also send you suitable sample bottles.

Microbiology and water treatment

Without doubt, the most important consideration for drinking water treatment is the removal of micro-organisms. A micro-organism is simply a very small creature, and they are divided into a number of subgroups (bacteria, viruses, fungi, protozoa and helminths). Micro-organisms from all of these subgroups can cause disease, in which case they are sometimes referred to as pathogens.

The main source of pathogenic micro-organisms in the aquatic environment is faeces (from humans or other warm blooded animals). The most common symptom of disease related to water supply is diarrhoea and/or vomiting (often referred to as gastroenteritis). This may be tolerable in healthy individuals, but depending on the infective agent and the dose, can progress to permanent and fatal organ failure (for example kidney failure in the case of children with E. coli 0157).

Given the huge variety of micro-organisms that may cause disease in water supplies, it isn't practical (or necessary) to test for all of them. Instead, we test for what are known as indicator organisms. The assumption is that any measure you take that kills *indicator organisms* will also kill any other pathogens that may be in the water. The two most commonly used in the UK are E. coli, and enterococci. Indicator organisms are discussed in more detail in the box below.

Box 3.1. Indicator organisms

The necessary characteristics of indicator organisms are basically that they are similar as possible to micro-organisms that may cause disease. Specifically:

• Present in large numbers in the faeces of warm blooded animals.
• Similarly likely or unlikely to die off naturally in the environment.
• Not otherwise present in watercourses.
• Similarly resistant/vulnerable to treatment.

• Not dangerous to lab technicians.
• Live longer than pathogenic micro-organisms.

As you can imagine, there is no such thing as an 'ideal' indicator organism, and two are commonly used. One is E. coli; the infectious dose (the number of bacteria necessary to produce an infection) is extremely low compared to many other bacteria. Enterococci are also tested for because they persist longer in the natural environment than many faecal bacteria and are more resistant to treatment. Both are therefore a conservative indicator of microbiological contamination. The limit for both these indicator organisms is 0/100ml; this reflects the fact that the presence of even a single indicator organism means that other potentially harmful micro-organisms might also be present. As you will see from table 3.4 other microbiological tests are also used, and the difference between results for these various micro-organisms is on occasion used by public health experts to help ascertain likely sources of contamination.

Whilst the use of indicator organisms is a reasonable approach when sampling water supplies, unfortunately there are some potentially harmful micro-organisms that are more resistant to certain types of treatment than the indicator organisms. The most relevant of these in the UK are *Cryptosporidium* and *Giardia*. These are protozoa, and are fairly resistant to disinfection because they form cysts; a protective shell around the micro-organism that isn't easily penetrated by chlorine. Both are common in surface water owing to the presence of animal faeces. Whilst most of us are fairly resistant to both parasites at low levels, both can be very dangerous to immune-suppressed individuals. Risks increase after heavy rainfall, since this causes agricultural slurry run off into surface water and hence increased levels of the protozoan in water supplies. Younger animals also have much higher Cryptosporidium levels in their faeces than older ones.

They are not routinely tested for in private water supplies, and as with all micro-organisms source protection measures (as described in Chapter Two) to prevent them getting into your supply is preferable to removing them once they are there. Both can be removed by filtration. Specific regulations govern precautions taken for Cryptosporidium in mains water supplies.

If my water is so dangerous, why am I not ill yet?

It is extremely common when investigating private water supplies to find dead animals floating in storage tanks, thick layers of muddy sediment in the bottom, and lots of green slime everywhere. It is a source of continuing amazement just how few people appear to suffer ill effects from their water supplies given the levels and frequency of microbiological contamination that occur, and the enormous numbers of private water supplies that are completely untreated. Even if a water supply looks clear, if a test result comes back as failing for microbiology, it is difficult for an Environmental Health professional to convincingly argue that the water supply is dangerous and that work is needed on it, when everyone associated with the supply claims that they've been drinking it for decades and never suffered any harm from it.

There are various reasons why this apparent paradox emerges. Firstly, you may have built up some immunity to common waterborne pathogens. However, this will not be the case for your visitors, and you should also bear in mind that the effectiveness of your immune system will decline with age, and that the immune systems of young children are poorly developed.

Secondly, it should also be remembered that indicator organisms are exactly that; they are not necessarily micro-organisms that make you ill, but their presence indicates that insufficient safeguards are in place to prevent water contaminated with faeces from coming out of your tap. You are therefore vulnerable to simple changes in the environment (such as heavy rain, or a change in the livestock grazing the catchment) that change the nature or the amounts of micro-organisms in your water supply.

Waterborne disease in the UK is vastly under-reported, since most infections result in diarrhoea that clears up within a few days, most people will not go to their GP. Many sufferers will attribute the infection to causes other than their water supply. The GP will often not investigate the micro-organism responsible, and unless large numbers of people are infected, it is highly unlikely that a public health expert would be called in to investigate what the link was between cases.

In addition to the above microbiological considerations, it should also be remembered that most of the chemical parameters that are dangerous in water supplies cause chronic problems rather than acute ones; not having had vomiting or diarrhoea from your water says nothing about whether you are going to suffer long term brain damage or increase your cancer risk.

What next?

If you follow the advice of this chapter, you will by now have a list of works needed on your supply, based on risk assessment and/or the checklist of table 3.3. You will also have some results from water sampling.

Bear in mind the limited relevance of microbiological testing and what the test results do and don't signify; if the results show the source has a relatively low level of microbiological contamination (for example <10 coliform/100mls), this is not necessarily a sign that the water is relatively safe. Unless you have implemented/installed source control measures and a treatment system, the results may simply be luck. It says little about the quality of the water under different environmental conditions (for example after rainfall, or when there is a rotting sheep carcass in the stream, or when the local farmer turns over-wintered cattle out into the field containing your water source). For microbiological parameters, risk assessment and/or the checklist of table 3.3 is more important than water sampling.

Whilst the values of microbiological parameters are extremely variable, chemical contamination is generally more constant, and the usual action in this instance is water treatment, as discussed in the next chapter. It is generally a good idea to repeat chemical tests once treatment has been installed, in order to verify that the measures taken have been sufficient.

Summary
Whilst single households on private supplies are virtually exempt from regulation (except in Scotland), your risk of getting ill from your water is considerably higher than from a public supply. The checklist in table 3.3 can be used as a starting point when considering how to improve your supply. Taking measures indicated by this checklist and getting your water tested for a handful of parameters as suggested in table 3.4 will considerably reduce your risk.

Further reading
Householders Guide to Private Water Supplies, Foundation for Water Research, 2010. Booklet particularly focusing on the regulatory side of private water supplies.

Small Water Supplies. A practical Guide, Clapham, Spon Press, 2004. Aimed at Environmental Health practitioners, contains lots of technical detail on health aspects of private water supplies.

WHO Guidelines for drinking-water quality, fourth edition, 2011. Downloadable from WHO website: www.who.int – the worldwide definitive (500 page) document on water quality.

Chapter Four

Cleaning water – practicalities

Filter types for removing solids – killing micro-organisms – removing micro-organisms by filtration – removing dissolved materials – mains and bottled water

As discussed in Chapter Three, for microbiological contamination the best starting point is source control and measures relating to the structures installed around the source. However, additional treatment will be necessary for micro-organisms and other contaminants. The starting point for this chapter is the findings of the water sampling you have undertaken.

Each of the parameters tested for has a limit in the PWS Regulations, and whilst you may not be subject to the regulations, these limits represent sensible practice based on worldwide guidelines related to either health or acceptability.

At a basic level, water contaminants fall into three categories: solid things, living things and dissolved things, and this is generally the order in which a treatment system will be designed to remove them.

Solids: These are often as simple as bits of sand and soil. They may or may not do you any damage in and of themselves, but a soil particle can be home to millions of bacteria. Bits of sand and grit can also damage your plumbing system and appliances, impart colour to the water, and prevent certain types of treatment from working effectively.

Living organisms: These can cause serious illness, and can either be killed, or removed by filtration (since they are solids, albeit small ones, and they often live on the surface of larger solid particles). The presence of micro-organisms is undoubtedly the most significant health concern associated with private water supplies, as discussed in the previous chapter.

Dissolved materials: There are lots of metals, minerals, gases and man-made contaminants that can be present in water. Some are dangerous to health in the long term, others impart flavour (pleasant or otherwise), and others affect more general properties of water such as pH and hardness, levels of which can be too low as well as too high. Most dissolved materials (with the exception of nitrate and nitrite) cause chronic health effects as opposed to acute, so "it hasn't made me ill yet" is no indication of water safety. Some can also damage appliances. Dissolved materials are removed by a variety of chemical and physical processes, as discussed on page 89 onwards.

Step 1 – removing solids

The initial approach for removing solids is source protection; that is preventing solids getting into your water in the first place, as discussed in Chapter Two. The amount of solids present is indicated by turbidity, and the PWS Regulations specify a value of 1 NTU (which stands for nephelometric turbidity units, and relates to how much light is scattered by the solid particles in a water sample). Solid removal can consist of one or several processes; settlement, coagulation/flocculation and filtration.

Settlement

If a solid has a density very different to that of water, it will float or sink fairly quickly, but this isn't the case for many of the solids in water. Settlement times tend to be so long that it isn't worth constructing large settlement tanks or reservoirs specifically to settle out solids but the methods described in Chapter Two for getting your water out of the source should take advantage of any settlement that is occurring naturally.

Coagulation and flocculation

Many solids in water naturally have a negatively charged surface, and therefore tend to repel each other. Coagulation is a process whereby a chemical (usually aluminium or iron sulphate) is added to the water to reduce this surface charge. Flocculation (a process in which these particles clump together in what are known as flocs) then occurs, and these larger flocs are easier to remove (by settlement or flotation).

This process is also fairly effective at removing micro-organisms, as they are themselves solid, and are often also attached to solid particles. Whilst it is routinely used by the water industry for removal of solids, natural organic matter (NOM), colour and algae, coagulation and flocculation are rarely appropriate for domestic scale water supplies owing to the careful control of chemical levels necessary and the complications of disposing of the metal-containing sludge which settles out.

Filtration

Filtering water is the most common technique used to remove solids on a domestic scale. Filters remove particles much smaller than would be predicted if they acted simply as sieves (the physical processes involved are diffusion, interception and sedimentation, which have distinct meanings in physics that need not concern us here). Some also remove a significant proportion of micro-organisms.

The rate at which the filter will get blocked varies according to the size of the holes; a fine filter will block more quickly than a coarse one. To avoid this, it is often best to have a coarse filter to remove larger particles, followed by a much finer filter. Some filter units have graded pore sizes to achieve this effect, with larger pores at the inlet compared to the outlet.

The speed at which water passes through the filter will also affect how many solid particles get filtered out; if you push water through a filter very fast, the solid removal will not be as good as if the water passes through more slowly. All cause some obstruction to flow and this will influence where in your treatment system they are sited. Coarse filtration (for example by gravel) to remove the large debris (>5mm) that may be present in surface water supplies (such as leaves and twigs) is considered in Chapter Two as part of the inlet structure for the system; you do not want this debris to get as far as your pipes as it is extremely likely to cause blockages.

Types of filters commonly used on water supplies are:
• Sand filters.
• Steel coil filters.
• Polypropylene/polyester cartridges.
• Bag filters.
• Ceramic filters.

Of the filters above, all apart from sand filters fit in housing units (such as in figure 4.1) that are compact enough to fit in a utility room or somewhere close to where the water supply enters the house. A variety of types of filter can be used, depending on what needs removing from the water and they can also be plumbed in series (for example if several stages of treatment are required), or in parallel (if a high flow rate is required or to decrease the frequency at which cartridges require changing).

Carbon filters and reverse osmosis units can also be categorised as filters, but their primary purpose is to remove dissolved material rather than solids, and they are discussed on page 94 and page 95 respectively.

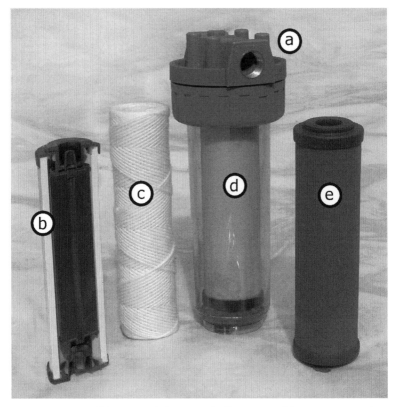

Figure 4.1. A range of filters available for (a) standard housing unit (b) cutaway version of a combined ceramic and carbon filter. The outer (white) region is ceramic and removes solids whilst the carbon core removes metals, pesticides etc by adsorption (c) string filter (d) ceramic filter (e) carbon filter (courtesy of Fairey Ceramics).

Sand filters

There are various types of sand filter, named according the speed at which the water flows through it, and the direction in which it flows. The most appropriate for domestic use is the slow sand filter. Two other types of sand filter used in the water industry but less suited to domestic use are rapid gravity sand filters (which remove 'floc' particles, described on page 78) and upflow sand filters.

A slow sand filter is illustrated in figure 4.2. Water enters at the top, at a rate controlled by a ball valve, which is positioned to allow a 30cm layer of standing water above the sand. The water passes down through a 0.5-1.5m layer of filter sand. Solids are removed by filtration, but an important

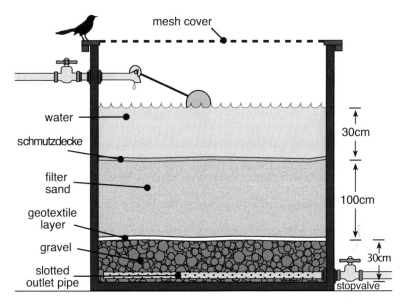

Figure 4.2. Slow sand filter.

additional characteristic of slow sand filters is that a population of bacteria and microscopic plants (known as a schmutzdecke), forms a fine mat in the top few centimetres of sand, and this mat of micro-organisms acts as a filter itself.

Micro-organisms living in this layer will also feed on pathogens in the water, and UV from sunlight will also kill pathogens. A layer of geotextile stops the sand mixing with a gravel layer underneath. Water flows freely through the gravel into a slotted pipe that takes the filtered water from the bottom of the tank, and this feeds into your water supply pipe.

The rate at which water passes through the sand filter is controlled by a valve on the outlet pipe which should be adjusted to a maximum of 100 litres/m^2/hour to provide good treatment (although up to three times this rate is used in mains water treatment works where 'multiple barriers' of additional treatment are used).

Special 'filter sand' should be used; grain sizes are in a narrow range (0.15-0.3mm) and it is therefore less likely to block with 'fines' (the very small sand particles that are present in common sand). Slow sand filters are good for water that already has fairly low turbidity; they are not ideal for water with colloids in (for example very murky river water where the solids are evenly dispersed and don't settle out easily). The sand filter should be covered to prevent debris falling into it, but since the action of UV light is beneficial, the cover should be some kind of mesh screen rather than a solid lid, unless there are lots of trees nearby (which might result in falling leaves, bird droppings and other sources of contamination).

Steel coil filters

These consist of a coil of stainless steel (rather like the walking spring toy known as a 'slinky') in a housing unit such as in figure 4.1. The surface of the coil has narrow ridges that keep each turn of the coil slightly separate, so that there are small gaps for water to pass through. Various size ratings are available, and they are most often used as coarse prefilters prior to other treatment systems.

They need backwashing at intervals, but automated units are available incorporating a differential pressure switch that senses when the pressure across the filter drops and then backwashes accordingly. A steel coil filter is a straightforward, compact method of pre-filtering water prior to further treatment for pathogens and dissolved substances. The fact that they are easily backwashed gives steel coil filters a definite advantage over wound polypropylene/polyester filters or bag filters.

Polypropylene, polyester filters

An old style technology that is basically string wound around a tube. A variety of viscose, polypropylene, polyester and nylon filters are now available, and they may be woven, pleated or bonded (figure 4.1). They are particularly appropriate for use with supplies where turbidity levels are already low. For higher turbidity water a coarser pre-filter (either one of the approaches to inlet structures discussed on pages 51-54, or a steel coil filter) will extend the lifespan considerably. They are not back-washable so must be disposed of when flow rate decreases.

Bag filters

These are as simple as they sound; a bag with very finely graded holes. They fit into cartridge units similar to those in figure 4.1. They are available in a variety of sizes (1 – 1000µm typically). They are expensive, and as with polypropylene filters, they cannot be backwashed so a steel coil filter may well be preferable.

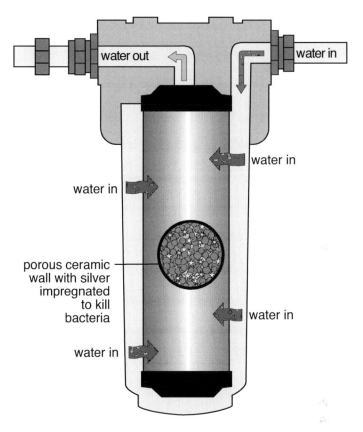

Figure 4.3. Ceramic filter. Water passes into the filter and solids are removed, including most micro-organisms. The filter is impregnated with silver which kills the micro-organisms that get trapped in it.

Ceramic filters

Ceramic filters (also known as ceramic candles, owing to their shape) are an extremely good way of removing very small suspended solids including micro-organisms (figure 4.3). Removal is quoted as >99.99% for particles of 0.9 μm, and >99.9% for particles 0.5-0.8 μm by a leading UK brand. This level of filtration provides good protection against micro-organisms, particularly in filters which are impregnated with silver (which kills micro-organisms trapped in the filter).

Owing to the small pore size, a pre-filter may be required to prevent the ceramic filter blocking too quickly. The filter is housed in the same standard unit already mentioned (figure 4.1) and can be removed from the housing

Figure 4.4. Two models of a counter top ceramic water filter. The model on the right has a much higher capacity and is shown split into its component parts; the ceramic candle is inside one section (providing the reservoir for untreated water), which sits on top of the other section (the reservoir for clean water). The model on the left functions in an identical way, and may be more in keeping with kitchen aesthetics (courtesy of Future Water Filters (left) and Royal Doulton (right)).

for cleaning. Cleaning is required when the flow rate slows down; indicating that solids are blocking up the external pores. These solids can be removed with a clean scrubbing brush or scouring pad, making sure that you do not contaminate the inside of the cartridge with unfiltered water. The filter will need replacing every few years as you will gradually scrub off the surface of the ceramic. In hard water areas, ceramic filters may suffer from premature blockage due to scale precipitation.

The case study on *Undergrowth Housing Co-op* in the box on page 93 indicates a typical installation.

Counter top ceramic filters are also available; rather like jug filters, these have a reservoir for the untreated water above the filter, and another below for the clean water. Since they require no plumbing they are good for temporary installations, or where water pressure is too low for a whole house treatment system (figure 4.4).

Step 2 – removing or killing the living organisms

Micro-organisms can either be removed by filtration, or killed. The latter can be achieved with chemicals, ultraviolet light or heat treatment. Many earlier stages of water treatment (including coarse filtration and storage) also help to reduce levels of micro-organisms. The degree to which a technique is effective against micro-organisms is usually expressed as a given number of 'log reductions' as discussed in box 4.1.

Box 4.1. Log reduction

Pathogen removal is often specified both in terms of the percentage removal and as a number of 'log reductions'. The relationship between these two expressions is shown in the table 4.1. The removal is also specific for a given named pathogen, because of their varying size and physical properties.

Log reduction	Percentage removal
1	90
2	99
4	99.99
7	99.99999

Table 4.1. The relationship between pathogen removals expressed as logarithms and percentages.

Whilst these percentage removals sound impressive, bear in mind that the limit for micro-organisms in drinking water is 0/100ml, and it wouldn't be impossible for a litre of untreated water to contain 1 million bacteria if an animal had died in your water supply (for example in an uncovered header tank). A 4 log removal (99.99%) would therefore result in there still being 1000 bacteria per 100ml water; more than enough to cause illness.

This emphasises how crucially important source control is; UV for example are typically rated to give a 4 log removal of micro-organisms and are usually regarded as a robust and appropriate treatment system. Any treatment system can therefore be overwhelmed, and their use should not be seen as a guarantee of safety, or a substitute for source control.

Filtration to remove micro-organisms

Micro-organisms are solids, albeit very small ones. Consequently, most types of micro-organisms can be removed by filtration, as long as the pore size in the filter is small enough. Even filters with large pore sizes will reduce levels of micro-organisms, partly because micro-organisms are often attached to soil particles, and partly because of the mechanisms by which filters operate.

Figure 4.5 shows the micro-organism removal profiles of various filter types. As you can see, slow sand filters and ceramic filters are both relatively effective at removing micro-organisms. Performance varies between studies, but a 3-4 log reduction in total coliforms is often quoted for slow sand filters. A leading brand of ceramic filters quotes a 4 log removal. Since micro-organisms that are trapped in the filter can breed and contaminate water passing through, most ceramic filters are impregnated with silver to provide a dual action of filtration and poisoning.

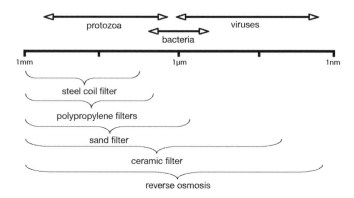

Figure 4.5. Filtration to remove micro-organisms. The pore size of various filter types affects their ability to remove micro-organisms. See text for the merits of different filter types.

Killing micro-organisms

Killing micro-organisms is generally achieved by damaging their genetic material so they cannot reproduce (UV), destroying them physically (boiling), or poisoning them with low concentrations of chemicals (chlorine).

Ultra violet light

UV is generally considered preferable to ceramic filters for microbiological treatment of private water supplies. A UV purification system for water is shown in figure 4.6. It consists of a UV light (shaped like the fluorescent tube in a strip-light) in a quartz sleeve, which the water flows past. UV light damages genetic material, so that the micro-organism cannot reproduce.

The sensitivity of micro-organisms to UV light varies according to the thickness of the cell wall; all common bacteria and viruses are effectively treated with the UV light intensities in commercially available treatment units (4 log reduction is often quoted), and UV is at least partially effective against Cryptosporidium and Giardia.

Water must be colourless for treatment to be effective (any particles or colour will cast shadows that could shield micro-organisms from the UV source) so prior filtration (to 5μm) is needed, and the flow rate through must be controlled to ensure adequate contact time. Low pH waters where iron staining occurs may well need additional pre-treatment to ensure that UV transmission through the water is sufficient.

Unlike chemical poisons, UV does not provide any residual disinfection,

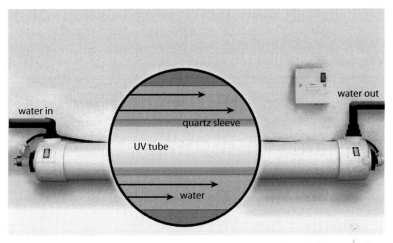

Figure 4.6. UV disinfection system. Water enters the sleeve around the UV bulb at one end and leaves at the other. The photo shows the bulb protruding slightly at the left end; the cutaway section shows what the inner sleeve and bulb would look like.

so any micro-organisms that get into the water afterwards can multiply. In terms of sustainability, UV treatment is not ideal since it requires electricity. The bulbs need replacing regularly as their brightness decreases over time, and need disposing of responsibly since, like other fluorescent tube lights, they contain mercury vapour.

The UV unit should be put somewhere convenient as you will need to check it regularly, and you should consider using one that alerts you if the bulb blows or the light intensity decreases; the major danger with relying on UV treatment is that when the bulb fails, it provides no treatment at all.

In addition to changing the bulb, the quartz sleeve that the UV bulb is contained in needs cleaning at intervals recommended by the manufacturer (for example once every six months); particularly in low pH waters where metal build up can occur, or in hard water areas where scale may form on the sleeve. Ease of sleeve removal for cleaning varies considerably between models; having a spare sleeve in case you break one whilst cleaning it is a sensible precaution.

Chlorine, bromine, iodine

The feature of chlorination which makes it the most widespread method for treatment of mains water supplies is that it provides residual disinfection after treatment so micro-organism levels will remain low during prolonged water storage and during transmission through the mains water network. However, chlorine is poisonous to humans at high concentrations, and is complicated to use:

- It reacts with other contaminants in the water to form unpalatable (and potentially poisonous) substances known as disinfection by-products (DBPs).
- The chemistry of chlorination means that varying doses are required to provide a suitable level of disinfection without taste impairment.
- The presence of other contaminants and variations in pH will also impair chlorine's effectiveness.
- It is ineffective against Cryptosporidium.

As a result, chlorination requires careful monitoring, and is rarely used for small private water supplies in the UK. Infusion pumps that constantly drip chlorine into water are available, but rely on a continuous flow rate of water of constant quality. If you are planning to use chlorination for disinfection of your supply you should obtain specialist advice from the equipment supplier. Other chemicals that are similar to chlorine are used on occasion; bromine is sometimes used in grey water systems, and iodine can be used for emergency water treatment, but neither is suited to regular use for water intended for drinking. Chloramines and chlorine dioxide may be used on large scale public supplies.

Silver

As discussed on page 83, many filter units are impregnated with silver, so they have a dual action. Large micro-organisms are filtered out, and then killed by the silver, and smaller ones (including viruses) can pass through the filter but get killed by exposure to the silver as they pass through. There is some concern that silver is only effective against certain types of bacteria. Silver is not appropriate as a domestic scale treatment technique, other than in impregnated filters.

Ozone

The ozone molecule (O_3 as opposed to O_2, which is the oxygen found common in air) is produced by subjecting O_2 to an electric field. Ozone is a very reactive molecule that will oxidise molecules on the cell membrane of micro-organisms, thereby killing them. Ozone can also remove organic contaminants from water supplies via oxidation. Ozone is increasingly used as an alternative to chlorination on municipal water supplies as it does not affect the taste of water. Just as with chlorine, the amount of ozone required varies according to the presence of other contaminants so working out a safe dose is complicated. Ozone is more effective than chlorine at removing Giardia and Cryptosporidium (though less so than filtration), but ozone provides no residual disinfection. Domestic ozonation units are available, but are expensive, have high power requirements and are not recommended.

Boiling and distilling water to remove micro-organisms

Boiling water for one minute will kill most micro-organisms and is a suitable treatment for water in an emergency situation. However, since it concentrates the water, levels of dissolved substances that can be harmful in the longer term can increase. Additionally, owing to the energy implications it is not a sustainable solution for regular use. Distillation can also be used to purify water; steam from boiling water is collected and condensed, and passed through a carbon filter. Since minerals in water are what cause it to taste good, distilled water is not very palatable. Solar distillation is also possible, and kits are available – designed for use in remote areas, or on boats.

Step 3 – controlling commonly occurring dissolved substances

For many private water supplies, removing solids and treatment for micro-organisms will be sufficient. However, as shown in table 3.5 (on page 70), there are lots of other potential contaminants. The most common in small private water supplies are discussed below. Many others are discussed on www.privatewatersupplies.gov.uk and the worldwide definitive guidance document from the World Health Organisation includes discussion of how limits have been arrived at for all controlled parameters.

Hydrogen ion content

Hydrogen ions (H^+) are responsible for acidity, which is measured on a logarithmic scale in pH units. A pH of 7 is said to be neutral. Lower than 7 is acid (lots of H^+ ions), higher is alkali (fewer H^+ ions), and the limits in the Private Water Supply Regulations are a minimum of 6.5 and a maximum of 9.5. The pH itself is not bad for you, but it affects what may be dissolved in the water, and both high and low pH can be extremely corrosive to your plumbing system.

Staining of porcelain or strange colours (particularly in the kettle) are often seen in existing supplies where the pH is low. Water with a low pH is likely to dissolve metals such as iron and manganese, and this is a common problem with upland surface water sources, which will not have passed through carbonate-containing rocks that would neutralise the acidity.

Sometimes organic acids from the breakdown of vegetation can also cause source water to be acid. pH can be raised (made more neutral) by passing the water through a filter that contains dolomite (limestone), which dissolves into the water and imparts alkalinity. The dolomite filter can be a sealed unit (that looks a bit like a SCUBA divers tank), or in a separate filter rather like a slow sand filter (as used in the Undergrowth Housing Coop case study on page 93.

Other compounds (for example magnesium oxide) can also be used, although some systems are prone to sedimentation and are not appropriate for irregular use (for example in holiday homes); seek advice from the manufacturer. Even in instances where low pH does not result in other parameters falling outside the limits in the Regulations, it should be neutralised in order to prevent corrosion of your plumbing system.

Conductivity

Given the sheer number of things that could be dissolved in water, you would need to perform lots of different tests on a water sample if you wanted to know exactly what was in it. Luckily, a fairly non-specific test is available to give an initial idea of whether there is anything in the water that is of concern. Conductivity (measured in micro Siemens/cm, with a limit in the UK of 2500) is a measure of the total concentrations of positively and negatively charged ions in the water. Since lots of contaminants can exist as ions, conductivity is a good general test for the presence of contamination.

Pure H_2O has low conductivity, whereas water with lots of ions in has high conductivity. Most ions likely to be present (such as sodium, potassium and chloride) are perfectly safe at levels that cause high conductivity, but more harmful ions such as arsenic, lead and cadmium could also be responsible for a high conductivity reading. Consequently if your water has a high reading you will need to have tests carried out on other parameters to find out what is causing it. A high conductivity reading could also be an indicator of saline intrusion into your groundwater supply, in which case you may well have to find an alternative water source.

Lead

The limit for lead in private supplies is 25µg/l until 2014, and 10µg/l thereafter (the progressive decrease in the limit is in order to give countries time to comply with what in some cases will be a challenging regulation). Lead can be from your plumbing or (more rarely) due to local geology. If your water contains lead due to the local geology (a local history of lead mining is an obvious indicator!), you should check the pH of the water; acidic soft water will dissolve lead, so you may be able to reduce lead levels sufficiently simply by neutralising the water. Ion exchange media (see page 95) can also remove lead in a private supply. When consumed over a period of years, lead is probably the most dangerous dissolved material that may be present in a private water supply as it builds up in the body and causes brain damage. Lead from plumbing systems is considered on page 102.

Nitrate and nitrite

Nitrates from fertilisers have found their way into our surface water and groundwater in many areas following years of intensive agriculture. Nitrate is also a breakdown product of ammonium, so can occur in waters that have been contaminated with sewage (so if your water test comes back with high levels of nitrate, you should ensure that the source isn't sewage).

Nitrite is formed when nitrate breaks down, and because the toxicity of both occurs via the same mechanism, the water quality limits are based on a combination of the two parameters. Unusually for dissolved substances, they may cause acute toxicity; there is a disputed link with 'blue baby syndrome' (methaemoglobinaemia) in bottle fed infants.

The simplest way to remove nitrate and nitrite on a domestic scale is with a nitrate specific anion exchange resin; water is passed through a unit that swaps the nitrate ions for chloride. Many of these resins preferentially remove sulphate (as opposed to nitrate), thereby reducing their nitrate removal capacity. Ensure that you get one that is specifically aimed at nitrate removal. They are available as standard sized cartridges. The EU nitrates directive and the UK implementation of it (a response to concerns over the environmental and health impacts of nitrates in water) has led to the classification of nitrate vulnerable zones (NVZ). Over time this should help to limit the problem of these contaminants.

Taste and odour issues

Qualitative assessments of taste and odour can indicate whether more specific tests need to be carried out. Solid organic material (algae, moulds, bacteria and breakdown products from plant material) can result in an unpleasant taste or odour, but will have largely been removed from the water by filtration. There are also some dissolved organic breakdown products that can cause taste problems. These are generally removed with a carbon filter (see page 94). A 'rotten eggs' smell in a well or borehole is a distinctly different problem. It is generally due to the presence of particular sulphur metabolising bacteria; shock chlorination (see page 134) is recommended. There are also a host of metals that cause taste issues in water supplies, as discussed below.

Iron and manganese

The limits on iron and manganese in the Regulations are based on amounts that impair taste or cause staining (of porcelain, and of laundry), and are far below those that cause health problems. Both metals can exist in two forms – one soluble, the other insoluble – and the most common removal techniques are designed to oxidise the metal so that it changes into its insoluble form.

This can occur naturally where the water becomes aerated (for example where the water is mixed, agitated or exposed to the air, such as at a spring where groundwater becomes surface water). The pH of water dramatically affects iron and manganese content; if your water is acid, simply neutralising it will cause iron (and to a lesser extent manganese) to turn into its insoluble form, after which it can be removed by filtration (as demonstrated in the case study of Undergrowth Housing Coop on the following page). Alternative approaches involve oxidising the metals either by aeration or by chlorination, followed by filtration, or a manganese dioxide filter.

Ammonium

Ammonium contamination of a private water supply is generally due to the presence of sewage (human or animal), so your first action should be to check the microbiological test results. If your water tests fail for both coliforms and ammonium, you should control the contamination at source. Ammonium from other sources is unusual, but can be removed by oxidation (which can be achieved using aeration). Since the oxidation of ammonium produces nitrates, you may well need to control nitrate levels in this instance. On a large scale ammonium is removed using chlorine, but this results in the formation of chloramines which in turn can cause taste and health problems.

Box 4.2. Case study – Undergrowth Housing Co-op

Undergrowth Housing Co-op is in a remote location in mid-Wales. It is 220 metres above sea level, and both mains electricity and water are more than two miles away. A stream runs past the house, and a small reservoir has been built that is used to run a hydro-electric turbine. The reservoir is 11 metres above the level of the house; this is not enough to provide sufficient water pressure for the household water supply once a cleaning stage has been included, unless a pump is installed. Instead, water is taken from the stream that feeds the reservoir at a point 18 metres above the level of the house. When analysed, this water had a low pH and consequently high levels of several metals (see table 4.2), in addition to high levels of coliforms and solids (as expected for upland surface watercourses).

Contaminant	Permitted levels	Before treatment	After treatment
pH	6.5-9.5	6.04	7.2
Manganese (µg/l)	50	116	7
Aluminium (µg/l)	200	290	65
Iron (µg/l)	200	469	63

Table 4.2. Effects of water treatment at Undergrowth Housing Co-op.

The immediate action was to install a ceramic filter underneath the kitchen sink (to remove the solids and the coliforms). When the residents had more time, a slow sand filter and storage tanks were constructed 15 metres above the house, to provide solid removal and decrease the frequency at which the ceramic filter became blocked. To neutralise the water (and so reduce the metal levels), an additional filter tank containing dolomite chippings was installed. Since installation, the water quality has improved (see table 4.2) and all metals levels are well within limits.

Aluminium

Aluminium is used as part of the treatment process in public water supplies, but is rarely found naturally in water at significant levels. Links between aluminium and Alzheimer's disease have been researched for several decades, but even the Alzheimer's Association agree that no causal link has yet been demonstrated. Since aluminium from drinking water will normally comprise less than 5% of total dietary intake there are easier ways of reducing your aluminium intake if you are concerned about it, than by trying to remove it from your drinking water. Aluminium is present in tea bags, anti-perspirants and at high levels in many antacid tablets. If you do wish to remove it from drinking water, specialist cation exchange resins are available but you should also check the pH of the water as neutralising the water (see page 89) may well reduce aluminium levels sufficiently.

Radiation

Total exposure to radiation in the UK consists of approximately 11% from medical sources (X rays and so on), 2% from nuclear activities, and the remaining 87% being from naturally occurring sources of radon and uranium in the ground. There are large regional differences in levels of naturally occurring radiation. If you live in an at risk area – for example where physical barriers under the house or mechanical ventilation is recommended to control levels of radon gas – then a private water supply from groundwater may well bring considerably more radon into the house. Radon is sufficiently volatile that aerating the water (in a well ventilated area such as an outside storage tank) so that the radon vents to the air, will reduce the level in the water. Also, since the half life of radon is short (isotope dependent, but 2-4 days), it can be controlled by storage. It can also be removed by carbon filters. The other naturally occurring radioactive compound of concern is uranium, which can be removed by ion-exchange, although finding an alternative water source is strongly recommended. Removal of radioactive sources from water supplies is a specialist area and professional advice should be sought.

Treatment techniques for dissolved materials

As discussed above, various techniques can remove dissolved materials and the right choice of technique is specific to the contaminants. If you are interested in treatment techniques *per se*, for example, if you have a range of contaminants to remove and are thinking about 'catch all' type systems, table 4.3 provides a basic guide to what can be removed by each one. Techniques for solid and micro-organism removal were discussed earlier, key features of techniques for dissolved materials are discussed below.

Activated carbon filters

Activated carbon filters trap contaminants in the filter by a physical process known as *adsorption* (which is not the same as *absorption* with a 'b'). This means that dissolved substances that would otherwise pass through the filter will bind to the carbon and so be removed from the water. They are called 'activated' carbon filters because of the way in which the filter medium is produced, resulting in a very high surface area per gram of material (500-1500m^2/gram) (rather like that in a honeycomb), providing a lot of binding sites for materials to be adsorbed on to.

There are two main types of activated carbon filter; granular activated carbon (GAC) and block carbon, and on private water supplies they are most often used to remove pesticides and herbicides. They are also effective at removing chlorine and improving taste, so they are often used in jug filters designed for use with mains water supplies. Block carbon filters have a more compressed structure, but require higher water pressures and pre-filtration to prevent premature blockage.

Activated carbon filters must be regularly replaced (frequency varies according to the manufacturer): once the binding sites on the surface of the carbon molecules are used up the filter provides no treatment, and can actually start releasing the contaminants back into the water again. Additionally, because they trap organic material, the filter itself provides an ideal breeding ground for micro-organisms so there is a risk that they can make the microbiological quality of water considerably worse.

It isn't practical to clean carbon filters; they must be thrown away after use, and there is no visual sign that the filter is no longer functioning properly. The most common types of carbon filter are available to fit a standard filter housing that is positioned under your sink (as illustrated in figure 4.1), and they often incorporate a small filter on their outlet to trap fines of carbon that may be released from the cartridge.

Ion exchange resins

Many dissolved contaminants can be controlled using a technique called ion exchange. A variety of units are available depending on the parameter to be controlled, but the principles behind them are the same. The ions (charged elements or compounds) that are impurities in the water are replaced by ions from the ion exchange resin. There are two basic types of ion exchange resin; cation exchange resins which exchange positively charged ions in the water for H^+ ions in the resin, and anion exchange resins which exchange negatively charged ions in the water for OH^- (or sometimes other anions) in the resin. If you need an ion exchange resin, you should make sure that the one you purchase is capable of controlling the parameters you are concerned with; ion exchange resins don't all do the same thing.

Reverse osmosis and membrane treatment

Filtration processes that rely on pushing water through a membrane are categorised according to the size of the pores in the membrane. Ultrafiltration, nanofiltration and reverse osmosis (in decreasing order of pore size) are all used for mains water supplies. In a reverse osmosis membrane (figure 4.7) the pores are less than $0.001\mu m$.

Treatment	Comments
Removal of solids	
Coarse gravel filter	For removing large (>5mm) debris. Best incorporated as part of the inlet structure (in order to prevent pipes blocking).
Settlement	Useful initial stage, but large storage volumes necessary.
Coagulation and flocculation	Difficult to control. By-products difficult to dispose of. Not suited to domestic scale.
Sand filtration	Effective and simple removal of solids and most micro-organisms.
Ceramic filter	Effective and simple removal of solids and micro-organisms, but may require preliminary coarse filtration to avoid filter blocking prematurely.
Steel coil filter	Compact (under-sink unit) and suitable for use as a pre-filter.
Polypropylene, polyester, viscose, nylon filters	Available as woven, pleated or bonded cartridges. Useful as a pre-filter (for example to a ceramic filter), but not suited to water with high turbidity (for example surface water), when an additional pre-filter would be required.
Bag filter	Effective removal of solids, available in a wide range of pore sizes, but cannot be backwashed.
Treatment	Comments
Removal of living organisms	
Chlorine	Good disinfectant, but can react with organic compounds to produce carcinogenic disinfection by-products. Difficult to control dose if levels of organic contaminants vary. Imparts undesirable taste, but provides residual disinfection, so allows storage.
Bromine	Not used for potable water owing to its toxicity. Sometimes used in grey water systems and swimming pool disinfection.
Iodine	Effective for emergency disinfection, but not recommended for long term use due to its adverse health effects.
Ultraviolet	Effective disinfection, via disrupting the DNA of micro-organisms. Water must be clear for UV to be effective. No residual disinfection.
Ozone	Dangerous by-products formed during treatment, not suited to small scale use.
Ceramic candle with silver	Ceramic provides filtration, with trapped micro-organisms killed by silver. No residual disinfection.
Boiling	Suitable for short-term use only. Energy intensive and often makes water unpalatable. Concentrates dissolved ions.
Distillation	Suitable for short-term use only. Energy intensive and often makes water unpalatable. Removes minerals from water.

Table 4.3. (Above and opposite). Techniques for cleaning water and the contaminants that they are designed to remove.

Treatment	Comments
Removal of dissolved materials	
Carbon filter	Remove halogenated organic compounds and chlorine. Used for taste and odour issues. Effective against a wide range of solvents and industrial chemicals. Often combined with ceramic filters for good all round treatment.
Ion exchange resins: softeners	Remove calcium and magnesium and replace with sodium. Resultant water is good for washing but not for potable purposes due to high sodium.
Dolomite filters	Neutralise acidity; will facilitate removal of metals including lead, iron and manganese.
Ion exchange resins: other cations	Removal of heavy metals, such as lead, arsenic, barium.
Ion exchange resins: anions	Remove nitrate and replace it with chloride. Will also remove sulphate, but if sulphate levels are high nitrate removal may be inadequate.
Reverse osmosis (RO)	Expensive, wasteful of water, energy intensive. Pre-filters in RO units (ceramic and carbon) often result in good quality water with little additional benefit derived from the RO. Not suited to drinking in the long-term and often unpalatable.

Its original application was to make sea water drinkable, and in countries with a scarce fresh water resource it is widely used on a municipal scale. Domestic scale systems are bad environmentally; the pump providing the necessary water pressure to push the water through the membrane will be operating at around 50 bars of pressure. The waste water the unit generates can be 8 times the volume of the clean water it yields (the volume depends upon the initial salinity of the water).

To prevent the membrane getting blocked, the unit includes prefilters (often both carbon and ceramic), so the water undergoes a good deal of treatment before it even reaches the reverse osmosis part of the unit. Backwashing and chemical cleaning cycles are also necessary in order to remove the particles that accumulate on the membrane.

Water that has undergone reverse osmosis will be more acid than the source water (because the bicarbonate ions will have been removed), so may cause corrosion. Post treatment is also necessary; minerals must be added back to the water in order to make it more palatable. Water purified by reverse osmosis is NOT recommended for drinking by the Department of Health (it is very soft, and there is some concern over a link between soft water and heart disease).

Technique	Bacteria & viruses	Giardia & crypto cysts	Algae	Coarse solids	Suspended solids & turbidity	Aluminium	Ammonia	Arsenic	Lead	Iron and manganese	Nitrate	Pesticides, solvents, industrial compounds	Taste/ odour
Gravel filter/ screen			+	++	+	+				+			
Slow sand filtration	++	++	++	++	++	+				+			
Steel coil or bag filter				++	+								
Polypropylene/ polyester filter			+	+	++								
Chlorination	++		+				++						
UV	++	+	+										
Activated carbon			++	++	++							++	++
Ceramic filter	++	++	++	++	++			+	+				
Ion exchange				++			+	+			++		
Membranes	++	++	++	++	++	++		+		++	++	+	++

Table 4.4. Summary table of treatment techniques. + is partially effective. ++ is effective/preferred technique for this contaminant.

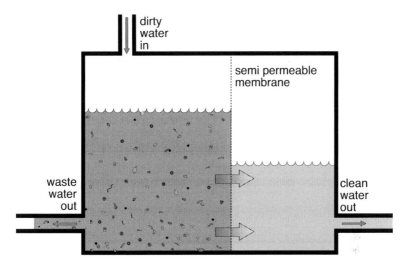

Figure 4.7. Reverse osmosis unit. If two solutions of different concentrations are separated by a membrane with tiny holes in it then water will tend to move from the more dilute solution to the more concentrated one. If pressure is applied to the side with the more concentrated solution, the direction of flow will be reversed. This causes pure water to move through the membrane, but keeps contaminants on the input side. The water on the input side therefore becomes more concentrated and is discharged as waste.

The only situation in which reverse osmosis is appropriate for private water supplies is when there is no other way of removing excess dissolved materials (for example if you have water with a high salt content). Whilst it isn't possible for reverse osmosis units to be water efficient, you could consider reusing the waste water from the unit for non-potable purposes such as toilet flushing.

Combined filters

The filter housings illustrated in figure 4.1 on page 80 can house treatment systems for most parameters singly, and can also be connected in series to remove multiple parameters. Some combine actions; for example, ceramic filters can incorporate an ion exchange medium (to remove heavy metals) and a carbon core (to remove organic compounds and chlorine), in addition to their basic function of solid and micro-organism removal. Whether or not you need any of these additional functions will depend on the results of your water tests; there is no point in having a filter that removes something that isn't there in the first place, and given that the different components of the filter will have different lifespans, it can result in a false sense of security.

Standards for water treatment techniques

As well as regulations on contaminants, there are a variety of certification schemes for water treatment units. Products should be labelled as NSF, WRc-NSF, or WRAS approved. BS EN standards also exist for many types of filter medium and treatment chemical, and the Drinking Water Inspectorate in the UK (DWI) has also produced definitive testing protocols for water treatment units.

Treatment units bearing any of these labels will have been tested against a rigorous set of protocols, the nature of which varies according to the type of treatment unit (for example for UV units they involve passing water containing bacteria through the unit, whereas for cartridge filters they involve passing water containing specific sizes of dust through the unit).

If your supply falls within the PWS Regulations, you must use water treatment products certified to these standards. You should also be aware that many water treatment products are not suitable for private water supplies; if you read the small print it may well include a statement that they are not to be used with 'water that is microbiologically unsafe or of an unknown quality'.

When to clean the water

As well as *how* to treat the water, you will need to consider *when* to treat it. At its simplest, this comes down to whether you treat water before storing it, or afterwards. Treatment techniques that operate slowly (such as sand filters) will need to incorporate storage buffers to help match supply and demand.

However, you should be aware that there is the potential for water to become contaminated during storage, so there is much to be said for treating the water immediately before it is used. Systems with multiple components may well have a system providing basic cleaning, after which the water is stored, before passing through another system within the house (for example a ceramic filter or a UV unit) within the house.

In practice, your choice will normally be governed by what treatment technique you are using (and how many stages), which in turn will depend on the contaminants and how robust a system is needed; where risks are high or the operational consequences of failure are major (for example with surface water sources serving temporary populations such as campsites or caravan parks), a 'multiple barrier' approach that incorporates several consecutive measures is justified.

Another major factor is whether water that has gone through partial treatment is of good enough quality to use for personal washing and bathing. As discussed on page 60, this should be of drinking water standards (because of the risk of ingestion). However, in households with no children and where everyone is aware of the risks, it may not be unreasonable to have (for example) the whole house served by a slow sand filter (to remove solids) and UV (to kill micro-organisms), but with removal of dissolved parameters achieved by an undersink filter at the kitchen tap, which is then the only drinking water source in the house.

A common reason for having two water qualities within the system is that the water pressure drop across cartridge filters is such that a whole house system would require a pump and/or additional storage. For filters relying on adsorption, the lifespan of the filter is also improved. This type of approach is particularly suited when there are dissolved contaminants that are dangerous over a lifetime of ingestion, but pose little or no instantaneous risk.

Mains water supplies

If you are concerned about water safety, your starting point should be to ensure that you are taking the appropriate measures to ensure that the water delivered through the mains stays clean. These measures are all contained within plumbing regulations, but you should check them in your home:

- Check that your header tank meets modern standards, including insulation of the tank and pipes (but not underneath the tank), overflow pipe, and a closely fitting lid. If your header tank is old enough to be made of galvanised steel it should be replaced.
- Ensure that outside taps have double check valves to prevent backflow (see page 187).
- Check for lead pipes (see page 102).
- Ensure that there is a check valve between your plumbing system and the washing machine and dishwasher. This removes the possibility of water that has been standing in the hoses and may have become tainted from flowing backwards into the rest of your plumbing system.

Filtering mains water and drinking bottled water are both becoming increasingly common. If you aren't happy drinking your tap water, you should start by deciding whether this is for reasons of taste or health. Most people's perception of water safety is closely related to taste, but within the typical ranges of mains water qualities this is generally an erroneous link.

Taste: This varies considerably across the UK, and according to the time of year, as water companies switch from one source to another and need to modify their treatment procedures to suit. Taste is a very subjective measure, but is often significantly improved by leaving the water in the fridge for a few hours; it may taste less 'flat' and any chlorine taste is less noticeable, so this should be your first tactic.

As discussed below, storing water in the fridge also decreases levels of disinfection by-products. If leaving it in the fridge doesn't do the trick, filters (either in a jug or plumbed in under the kitchen sink) can be used. These are usually carbon filters and their main action altering taste is the removal of chlorine (and therefore increase the risk of microbiological problems with the water). Carbon is also effective against MIB and geosmin; two breakdown products of organic material that are detectable at very low concentrations and make water taste 'musty'.

Health: The public water supply in the UK is sampled several million times a year, and in 2010 99.96% of those samples passed internationally approved tests. In contrast, routine monitoring of private water supplies suggests that over 80% fail similar tests. There have also been high profile incidents of bottled water being found to be contaminated; indeed bottled water can pose a particular risk since it doesn't generally incorporate residual disinfection. Nevertheless, you may be concerned about the effect on health of a few key parameters in mains water supplies.

Figure 4.8. An old lead water pipe entering a property, with the characteristic slight curve in the lead pipe (section near the ground), compared to the section of copper pipe (above and immediately below the tap) (courtesy of Metro Home Inspectors).

Lead

The limit for lead in public drinking water is 25µg/l reducing to 10µg/l in 2014. Lead was phased out for domestic plumbing around 1970, so if your plumbing is more recent, then it should not contain lead, but the most likely section that may not have been removed is where it comes in to the house (often under the kitchen sink). Check this section; lead is a dull grey that becomes shiny when you scratch the surface off with a screwdriver. It is also quite soft, so lead pipe is often slightly curved (figure 4.8). There is often a characteristic 'bulge' in the pipe just before a joint.

Guidance is available from the Drinking Water Inspectorate on action to take if you have lead pipes; generally if you replace the section you are responsible for (between the stop tap on the street and your house), the water company is obliged to replace the section of lead pipe that they are responsible for (the section between the stop tap on the street and the trunk main) if they have not already done so.

Grants may also be available from your Local Authority for lead pipe replacement. Water companies use both pH adjustment and phosphate addition to water supplies as methods for controlling lead levels where they know that there are a significant number of properties with lead water pipes. Raising the pH reduces the amount of lead that gets dissolved, whereas phosphate addition stabilises lead pipes by forming a layer of lead phosphate scale that prevents further lead dissolving into the water.

Fluoride

A number of water companies add fluoride to the public water supply at the request of the local health authority (artificially fluoridated water supplies about 10% of the UK population). Overwhelming evidence indicates that low levels of fluoride offer protection against tooth decay, but opponents of fluoridation would argue that most people get enough fluoride from their toothpaste and food to achieve these beneficial effects.

Studies have failed to find conclusive links between fluoride intake at normal levels and adverse health effects, although prolonged, artificially high intake has led to adverse effects on teeth and bone. If you are worried about fluoride you should start by removing other sources of fluoride from your diet (fluoride levels are high in seafood, toothpaste, tea and soft drinks).

If you wish to remove it from your water supply as well, you can install an activated alumina filter cartridge that fits in a standard housing (figure 4.1 on page 80). Many people object to fluoridation of water supplies on the basis of ethical concerns of what they regard to be widespread semi-compulsory medication. This is inconsistent when you consider that mains water treatment almost always including 'adding' substances to water as well as taking others out.

Chlorine and disinfection by-products

One of the drawbacks of using chlorination for treatment of mains water supplies is that it can react with organic substances in the water forming a range of compounds termed disinfection by-products (DBPs). Public concern relates more to the fact that you can taste these than to any health effects; the concentrations of DBPs that adversely affect health are far higher than those at which adverse taste is experienced and the risk to health is vanishingly small compared to the risks associated with inadequate disinfection.

However, if you are concerned you can ask the water company for copies of their test results, so you can compare them to the levels indicated in table 3.5 (see page 70). Storing water in the fridge significantly decreases concentrations of several DBPs, as well as improving taste. Whilst jug filters (containing activated carbon) can be used, the fact that these remove chlorine means that they provide an ideal breeding ground for bacteria. Some public water supplies now use chloramines (as opposed to chlorine) as it provides more robust disinfection and has a less noticeable taste. Unfortunately, they are also considerably more difficult to remove using domestic water filters.

Water hardness and softening water

As rain falls it dissolves carbon dioxide from the atmosphere and becomes a weak solution of carbonic acid. This slight acidity in the water means that it will dissolve minerals such as calcium and magnesium in the rocks that it passes through when it hits the ground. Groundwater supplies in areas with calcium or magnesium-rich geology, and surface water supplies fed by groundwater in these areas, are said to suffer from 'hard' water. Areas of the UK with hard water are illustrated in figure 4.9. Hard water increases the amount of soap required to produce lather (and so for some people can exacerbate eczema), and cause detergents to create scum. Dissolved calcium and magnesium carbonates in water can precipitate out as scale in pipes and appliances. A 1mm layer of scale can reduce efficiency of heat transfer by around 20%, so impairs the efficiency of immersion heating elements.

Water softening units are based on cation exchange; water is passed over a resin in which the calcium and magnesium ions that cause scale are replaced with sodium ions from the resin. The unit includes a tank containing salt water that automatically backwashes the resin periodically, to recharge it with sodium ions.

However, the high sodium content may cause cardiovascular disease; if you install a water softener on your mains supply you must also have a

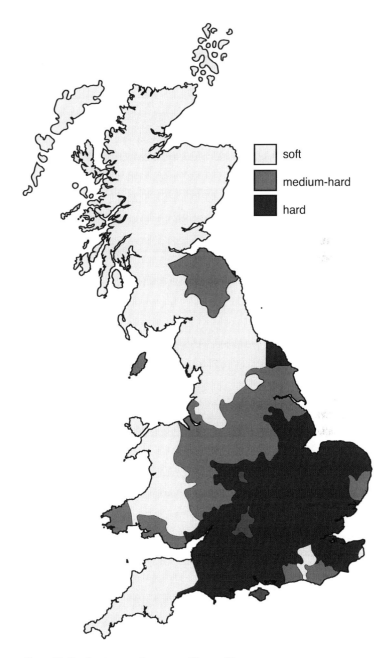

Figure 4.9. Hard water areas (courtesy of Avonsoft).

dedicated drinking water supply that has not been softened. Softened water is particularly dangerous for bottle fed infants, or people on low sodium diets. Also note that using any water treatment system that incorporates backwashing, regeneration, or produces some kind of wastewater discharge is 'notifiable'; you need to inform your local water company before installing the system.

Other types of unit (including magnetic, electrolytic and electronic) are known as 'scale inhibitors' or 'physical conditioners' and do not actually soften water. However a few studies have suggested that they reduce the likelihood of scale forming. Results with these units are variable; you should ensure you are offered a money back guarantee if you are purchasing a scale inhibitor.

Bottled water

'Nutritional information' labels on bottled water consist largely of data that is fairly irrelevant to health; the contents must comply with the same parameters that public and private supplies do. There is no health benefit to be derived from drinking mineral water, the main difference between bottled water and tap water is price; bottled water can easily be 3000 times more expensive and may have been shipped round the globe. The higher mineral contents of some bottled water makes them unsuited to infants and young children, and people with certain medical conditions, although manufacturers are not required to state this on the label. In most instances, if you need a bottle of water, the best thing to do is to fill one from your tap.

Summary

The way in which we clean water depends on the parameters we are trying to control, and these in turn will vary according to the end use of the water. The first stage therefore is always to get a sample of the water tested. The contaminants of concern are usually solids, pathogens and dissolved substances, and water is generally treated for each of these in turn. Filtration is often the main treatment for domestic supplies, and a range of filter types with various properties are available to remove a variety of contaminants. UV provides a useful additional safeguard against microbiological parameters. Together with the source control techniques and preventative measures discussed in Chapter Two, this 'multiple barrier' approach to maintaining water safety considerably reduces the risks arising from a temporary failure of any single part of the system. Types of treatment technique are indicated in table 4.3 and 4.4.

Further reading

Bottled Water. Understanding a Social Phenomenon, Ferrier, 2001.
Report for WWF summarising the environmental and social implications
of bottled water.

Water Filtration Manual (Oxfam Water Supply Scheme for Emergencies).
Downloadable from www.oxfam.org.uk

Water Treatment Manual (Oxfam Water Supply Scheme for Emergencies).
Downloadable from www.oxfam.org.uk

Websites

www.privatewatersupplies.gov.uk – guidance on private water supply
regulations for the different parts of the UK. Includes the definitive 600
page guidance document on compliance, and a more user-friendly version
for members of the public.

Drinking Water Inspectorate leaflets are available on the most common
areas of concern with mains water supplies. Downloadable from
www.dwi.gov.uk

Chapter Five

Fitting it all together

Storing water – moving water – the physics of flow - choosing pipes – choosing a pump – wind pumps – hydraulic rams – human powered pumps – electrical pumps – useful controls

Once you have found your water source, worked out how to get the water out and how to clean it, you need to think about getting it from the source to your house. You will want to do this in the most energy efficient way possible and without contaminating the water. You will need to choose pipes and fittings, and, if you need to move the water uphill or over some distance, you will need to choose a pump. Your water system will have the following components:

• An inlet structure (considered in Chapter Two).
• Storage.
• Pipes.
• An energy source.
• Control mechanisms.
• Cleaning processes (considered in Chapter Four).

The precise layout will vary between systems, particularly in terms of the location of the treatment and storage components.

All components in contact with water should be suitable for potable use; they will be marked as "WRAS approved material" and you can check on individual products using the WRAS fitting and materials directory.

Water storage

Water storage is needed to buffer differences between supply and demand, and to provide a backup in the case of failure of the source or pump. The storage also provides some treatment (by settlement of solids, and due to the tendency of pathogens to die off over time). Features of typical storage tanks are shown in figure 5.1.

Lightproof lockable lid: storing water in the dark decreases growth of algae and other micro-organisms. Preventing unauthorised access is good health and safety practice. For underground tanks, the lid may have to withstand loads.

Stop taps and drains: a means of isolating storage from the system makes draining, cleaning and maintenance easier.

Smoothing inlet: Opportunities for settlement should be maximised by ensuring that incoming water is to the bottom of the tank, rather than splashing onto the surface.

Tank material: plastic is most common, although metal is also used, particularly for larger tanks. Both are available in a wide variety of shapes and sizes. When buying tanks you should ensure that they are suitable for potable water, and if they are going to be above ground, that the plastic is UV stabilized.

Overflow: Whilst this may seem like a waste of water, it is important that you don't allow your stored water to stagnate over a period of weeks. You may be able to put this overflow water to good use, or return it to the water source (for example surface water) so that it continues through the hydrological cycle. The overflow should be vermin proof (for example incorporating a U bend), and the end should face downwards.

Siting: In practice, most private water supplies will have above ground storage, as it is cheaper and simpler. It does however result in higher water temperatures during summer (which increases bacterial growth). Structural stability needs to be considered when installing tanks underground; differential ground movement, resisting vehicle loadings, resisting ground pressures and water table fluctuations (both of which can cause the tank to deform) can all present problems. It is also not unheard of for tanks to pop out of the ground (if the water level is low and pressure from surrounding ground is high). For this reason, many underground tanks will have concrete surrounds.

How much to store

There are two issues to consider with regard to how much water to store. Firstly, your total water use and, secondly, peak demands. Typically, where the likelihood of the water supply drying up in the summer is reckoned to be low, storage volumes are 1000 litres, or enough for the household for 2 days, whichever is greater (you should have a good idea of how to calculate this from Chapter One).

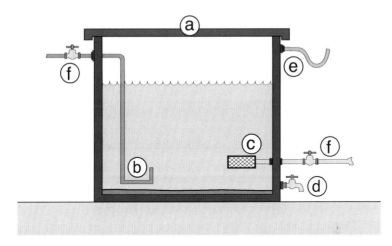

Figure 5.1. Water storage tank. (a) Lightproof lockable lid (b) smoothing inlet to maximise settlement (c) outlet with strainer (d) drain (e) vermin-proof overflow (f) stop tap.

However, if there is a chance that your water supply will dry up for a period, you will need to store enough to last for this time; it may well be difficult to know this in advance, and it may be worth having reserve storage tanks, or at least providing sufficient ground area during your initial installation works to make it easy to install them at a later date.

Ultra low water use households may have peaks in demand that easily double average daily use (such as for an occasional bath instead of a shower). If you have regular guests you should take these peak demands into account when deciding on a size of storage tank.

If your water use varies according to the season, you may choose to have reserve storage tanks that you only use for part of the year. These will need cleaning and disinfecting before being brought into use. If this extra demand is for non-potable water, such as garden use, consider having a separate store of water for this purpose.

If you are abstracting groundwater from a spring or a well, you may not need to include an additional storage tank, since these methods of accessing groundwater often include a storage reservoir; a household spring box can be built to the required size, or if you have a well, the cavity within it will act as storage. However, if you are pumping water from your groundwater source rather than relying on gravity, incorporating storage will increase the efficiency of the pump (as discussed below), and provide a buffer in the event of pump failure.

How much energy do you need to move your water?

Moving water takes energy, and you will need to calculate whether or not you can rely on gravity or will need a pump. There are three parts to this calculation:

1) Measuring the height difference.
2) Determining the necessary pressure and flow rates within the house, and the highest point at which these pressures and flow rates are required.
3) Calculating the pressure loss caused by other components of your system.

1) Measuring the height difference

You will need to measure the difference in height between your source and your supply point. This is known as the *static head*, is usually expressed in metres, and is the major determinant of energy requirements. When you measure this difference you should determine the level of a few landmarks along the way as well as the total head so that you can create a profile, which will allow you to choose an optimum position for your storage and treatment systems. There are several ways of measuring differences in height, two of which are described in the box opposite.

If you have a mains water supply, your water will be at around 2-3 bars of pressure (although this will vary according to the demands on the network at a given moment). Private water supplies usually don't need to run at pressures this high, and your pressure requirements will vary according to your appliances. Many water treatment systems will have minimum operating pressures (typically 0.5 bar for cartridges in standard housing units). An instantaneously heated electric shower will usually have a requirement for a minimum of 1 bar (but, given their energy inefficiency, you should try and avoid these anyway). If your hot water system includes a header tank, you should make sure that either a) the header tank is large enough for your biggest hot water use (the usual approach), or b) that the flow rate to the header tank is sufficient to meet the water demand. Some other appliances also have minimum pressure requirements.

In practice, your choices about pressure and flow will probably depend upon whether or not you are expecting to need a pump. If, based on your initial head calculations you are expecting to need a pump, then you should check with the pump supplier what pressures the pump will generate (although you can generally assume that even the smallest pump will provide water at over a bar of pressure).

Box 5.1. Measuring height differences

Figure 5.2. 'Dumpy' laser level used to measure the height difference between two points.

A laser level consists of a tripod stand with a laser light on the top that you project horizontally onto a staff held further up the slope (figure 5.2). The staff has markings on it so that you can record the height difference between the laser and the staff. This is repeated as many times as necessary until you get to the top of the slope. They are widely available, so you should be able to hire or borrow one.

A clinometer (figure 5.3) measures angles, and some basic trigonometry will enable you to turn an angle into a height difference. Take two posts of the same height and position one at the bottom of the slope and the other further up (as far away as you can comfortably see it). Place the clinometer on top of the first post and sight to the top of the second post, taking a note of the angle. Record the distance along the ground between the two posts, as well. Repeat this procedure by moving the posts until you get to the top of the slope. You can then use the sine rule to calculate the vertical distance between points (figure 5.4). Clinometers are much quicker for measuring big differences in height than laser levels. They can also measure angles below the horizontal, so are easier to use on undulating ground than laser levels. However, they require you to be able to accurately measure the distance along the ground.

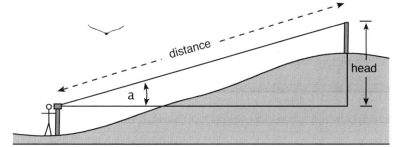

Figure 5.3. Measuring head using a clinometer and the sine rule. You will know the angle of the slope from the clinometer. The sine of this angle, multiplied by the distance between the two points along the ground will equal the head. For example, if the clinometer reading is 14°, and the distance along the ground is 35 metres, the head is sine 14° x 35 = 8.5 metres.

Figure 5.4. Clinometer, used to measure gradients.

3) Other sources of pressure losses

The total head which we discussed measuring on pages 112-113 will not necessarily be the difference in height between your water source and the points of water use in the house. If you have any storage or treatment components within the system that are open to the air, it is the height at this point rather than the height of the source that will be relevant in the head calculation, as shown in figure 5.5. You must take this into account when positioning treatment and storage tanks, and they are generally installed as high in the system as possible. In the unlikely event of the water pressure being too high due to a high head, pressure reducing valves can be used. These decrease the risk of pipe movement and bursting, and also improve water efficiency.

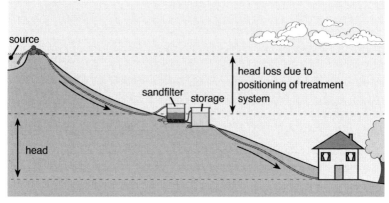

Figure 5.5. Head loss due to treatment system. In this example, the actual head of pressure available in the house is lower than the head difference between the source and the house, since the treatment system is positioned part of the way down the hill and is open to atmospheric pressure. To maximise the available head, the treatment system could be moved further up the hill. Within the house itself, the height above ground floor level that you have water using appliances will also need to be taken into account (but is not illustrated here).

When water in the system isn't moving, a pressure gauge will give a reading that is referred to as *static pressure*. If you then turn on a tap, the pressure will drop. The lower pressure that occurs when the water is flowing is termed *dynamic pressure* and it is this pressure that will determine how fast water will come out of a tap. The reason that the pressure drops is that energy is lost in the pipe and fittings.

The energy lost within the pipe is dependent on the pipe diameter and the flow rate, and can be calculated as discussed in the box on head loss. The amount of energy lost within bends and pipe fittings varies according to the type of fitting, and can be expressed as an equivalent length of straight pipe and then added to the losses you calculated for the pipe itself. An example of this calculation process is described in the box, along with some more of the physics that governs pressure and flow.

Box 5.2. Calculating head loss in water supply systems

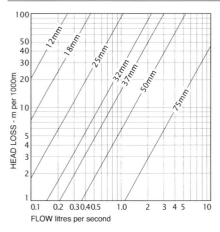

As discussed earlier, *head* is the vertical distance between two points. *Pressure* is the force that something exerts, and is proportional to the density of the water (which is constant) multiplied by the head. The SI units are pascals, but you will often see reference to bars, pounds per square inch, metres of water and atmospheres. Most of the time, we are more interested in dynamic pressure than in static pressure, and will need to calculate the difference between the two in order to specify a pump, or determine whether or not water will flow through your system under gravity alone.

Figure 5.6. This graph shows the amount of energy (expressed as head) lost for every kilometre of MDPE pipe. This varies according to the flow and pipe diameter. Similar graphs are available for pipes of different materials (courtesy of Practical Action, formerly the Intermediate Technology Development Group).

The amount of friction is dependent on the smoothness of the pipe wall, and the speed at which the water is flowing along the pipe; it varies according to pipe material and size. The graph in figure 5.6 shows the energy loss (expressed as vertical

Pipe bore (mm)	Equivalent pipe length (m) for fitting type		
	Elbow	Tee	Stop valve
12	0.5	0.6	4.0
20	0.8	1.0	4.3
25	1	1.5	5.6
32	1.4	2.0	6.0
40	1.7	2.5	7.9

Table 5.1. Typical equivalent pipe lengths for common pipe fittings (valid for copper, plastics, stainless steel). From BS6700.

head of water) for MDPE pipes of different diameters at different flow rates.

Similar graphs or tables exist for all pipe types. There will also be losses in the pipe due to turbulence caused by obstructions such as bends and taps. Table 5.1 shows an equivalent length of pipe so that you calculate these losses as well. If you don't want to do the maths, then you can usually assume that 25mm MDPE will be sufficient for single house applications where the supply pipe is less than about 300 metres long; head losses are low unless you are transporting water large distances, only have a very small pipe, or need a particularly high flow rate (for example for power showers).

Note: The mathematical relationship between water pressure and flow is given by Bernoulli's equation and is beyond the scope of this book. It can be used to calculate water pressure under dynamic conditions; for example when water is flowing at a known rate.

Pipes

Material	Comments
Polythene (polyethylene)	Flexible, rot resistant and fairly robust. Sold in coils up to 150 metres. If subjected to frost it will tend to burst at the joints rather than split the pipe, making repairs very easy. Two types are available; medium-density polythene (MDPE) and cross-linked polythene (PEX). Not suited to hot water. Standard sizes are 20, 25, 32, 40, 50, 63mm. Usually blue, (old pipe is black). Joints and other fittings are expensive.
Galvanised steel	Not recommended owing to corrosion problems.
uPVC	More commonly used for foul water drainage than potable water supply.
Polybutylene (PB)	Flexible and robust, commonly used within the house.
Copper	Expensive for long supply pipe runs with large diameter but is extremely suited to shorter pipe runs in houses. Connected by soldered joints (cheap) or compression fittings (more expensive). Will split if frozen.

Table 5.2. Common pipe materials and their features.

Water pipes suitable for carrying domestic water supplies come in a variety of materials, the most common of which are shown in figure 5.7 and compared in table 5.2.

Polythene MDPE pipe is by far the most common choice to move water between the source and the house, with copper or polybutylene used within the house as they are compatible with hot water systems. Composite pipe

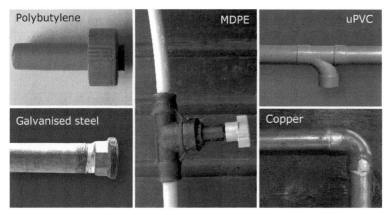

Figure 5.7. Pipe materials.

made of plastic with an impregnable metal layer is occasionally used for water mains where the ground is contaminated.

Whatever sort of pipe you use, check that it is suitable for potable water supplies so you don't get chemicals leaching into your water supply, and that the rated pressure is sufficient (for a private water supply it is unlikely that you will need anything other than the standard rating for MDPE, which is 12 bar). You should make sure you have your water tested (as described in the chapter on cleaning water), as corrosive water may influence your choice of pipe material (for example avoid copper if the water is corrosive). The type of plumbing fittings you use will be also be determined by your choice of pipe.

Laying pipe

Your water pipe will need to be protected from mechanical damage, ultraviolet light and frost. The best way of doing this is to bury it in a trench. Recommended depths in British Standards and Water Regulations are generally 75cm, but this is rarely done on private supply systems; 30cm is more common. You should make sure when laying the pipe that it is free from kinks, sudden changes in direction or height, and that there isn't anything sharp in the trench that could damage it. Don't pull the pipe completely straight, small bends should be left to allow movement should ground settlement occur.

The system should be checked for water tightness before backfilling the trench, and you should avoid burying any connections or taps; install a manhole at these points. At points where the pipe is going to be exposed to the air (for example in manholes) it should be insulated to prevent freezing, and this should be done during installation, so that the insulation can extend into the backfilled trench. If you choose to bury your water supply pipe, ensure that its route is detailed on a map, marked on the ground, or walk the route with a GPS device.

Given the labour and costs required for burying pipe you may prefer to run the risk of occasional freezing, or lay the pipe on the surface and insulate it instead (check the technical data to ensure the insulation you use is suited to outdoor use). In most instances 13mm thick nitrile rubber insulation is sufficient for a 25mm MDPE pipe.

Attention to detail is important when insulating pipes; better insulations are glued, some require additional waterproof external membranes, and many should have cable ties at intervals or vulnerable points to avoid any exposure of the pipe to air temperature. The point at which freezing is most likely to occur is around any float valves, so pay particular attention to them when insulating. Insulation should also be inspected annually to check its integrity.

Other measures of preventing freezing include: TRACE heating (electrical heating tapes placed between the pipe surface and the insulation); leaving a tap dripping; and 'frost stats'. None are really substitutes for good insulation and there are no convenient rules of thumb for what flow rates are required from dripping taps to prevent freezing. If you are going to resort to this, make sure it is not a tap that discharges to a sewage treatment system (in other words use an outside tap, or one that runs to surface water drainage). Obviously this is also a waste of water, but whether this is a significant problem depends on the nature of the water supply.

Pumps

By this stage, you will have got as far as choosing a type of pipe and how big it should be. You will also have measured the head in your system, and calculated head losses due to the pipe and fittings. The end result is the total head within the system. You can then take this value, and the amount of water you need to move (which you worked out in Chapter One) so that you can work out what energy source is most appropriate for your water system.

If the head in your system is sufficient to produce an adequate flow rate into your header tank and to all your water using appliances, you may not need to pump your water at all. However, many instances will require some sort of pump. Pumps are specified by referring to their output (in litres or cubic metres per unit time) versus the head that this water is pumped to, and this is either given in a graph or a table, as shown in table 5.3 (for a typical submersible pump for small scale applications). Principles and mechanisms of pumping are dealt with in the box below.

There are four main energy sources available to you that could power a pump; wind, water, human and electrical.

Head	Capacity	
(metres)	m³/hour	Litres/sec
10	4.8	1.25
15	4.2	1.17
20	3.6	1
25	2.9	0.8
30	1.8	0.5
35	0.6	0.17

Table 5.3. Pump output versus head for a typical submersible pump suited to domestic supplies.

Box 5.3. How pumps work

Pumps have two actions. Firstly, the pump must 'pull' water into itself and secondly, it must push water out of the other side. The relative amount of each of these actions leads to classification of pumps into those that *push* water (submersible pumps) and those that *pull* water (suction pumps), in other words whether the pump body is situated above the water (suction pump) or in the water (submersible pump).

The practical difference between the two types is the amount of head they can pump water to. Suction pumps cannot move water to a head higher than around 8 metres (they are limited by atmospheric pressure, so the theoretical maximum pumping head is 10.4 metres at sea level, but is lower in practice due to efficiency losses), whereas submersible pumps can be found operating in boreholes that are 70 metres deep. Suction pumps will not operate if the supply pipe contains air; the pipe needs to be filled with water in a process called 'priming'. Whilst most suction pumps are self-priming and can remove small amounts of air that may enter the pump during operation, you will generally need to supply the initial priming water.

Pumps can also be classified according to their mechanisms of action:

Piston pumps: This is the classical design used for hand operated pumps. On one stroke a plunger creates suction that draws water through a valve into a cylinder. On the return stroke the water is forced out through a second valve. If pumping to a large head the plunger needs to be at depth so that most of the action is via pushing rather than pulling.

Rotary, centrifugal and helical pumps: These have a rotating part called the impeller (like a propeller), which as it rotates tends to suck water into the pump and push it out of the other side. In helical rotor pumps the impeller consists of a spiral shaped rod. Centrifugal and helical pumps can operate to considerable heads, and can be combined in a multi-stage pump in high head situations (each stage is basically a pump with its own impeller, that pushes water up to the height of the next pump).

Jet pumps: Water travels through a narrow tube, which widens at the outlet end. As water flows into the wider section it creates a partial vacuum which sucks more water in to the pump.

Peristaltic pumps: A flexible tube is squeezed by rollers, thereby pushing the liquid through the pipe. They are not used for water supply pumps.

Wind and water powered pumps

In some circumstances, both wind and water can be used to power pumps directly, without converting the energy to electrical power first.

Wind pumps (figure 5.8) have been used for hundreds of years to pump water from wells and boreholes. They consist of a multi-bladed rotor, attached to a crank acting on a reciprocating piston. Clearly, they only pump when the wind is blowing, so you will either need to store enough water to see you through calm periods, or have a standby pump powered by another means.

Figure 5.8. Windpumps can be used to abstract water from wells and boreholes.
The spinning blades are attached to a crank mechanism which drives a piston pump.
Windpumps can move thousands of litres a day from depths of over 50 metres, with zero electricity cost, however their capital cost is high and they are best suited to large supplies or where electricity is not available.

An additional problem is that the mechanism will be damaged if you try to prevent the pump from pumping water when it is windy, so you will need to incorporate a facility for overflow from the storage tank, although preferably not one that will be wasted (but you should not connect the overflow back into the well/borehole directly as it will potentially have become contaminated by reaching the ground surface).

The capital cost of wind pumps is high (over £2000), so they will only be economic for large scale supplies where running costs would outweigh capital costs (in other words thousands of litres per day), or where there is no mains electricity supply available. In these instances, the wind resource should be estimated and used to specify the correct capacity of wind pump and a suitable storage volume (see Further reading for details).

Hydraulic rams (figure 5.9) are water powered pumps that use the kinetic energy in the incoming water to alternately open two valves and pump a proportion of the incoming water to a height (see figure 5.10 for a detailed explanation of how they work). The proportion pumped to height depends upon the head available in the incoming water, and the head to which the water must be pumped (table 5.4). They can therefore work very well in situations where surface water from a stream is being pumped to some distance above it.

Figure 5.9. (right) Hydraulic ram pump.

Figure 5.10. (below) Hydraulic ram operation a) pulse valve, b) delivery valve, c) delivery chamber, d) supply pipe, e) drive pipe. 1) Water flows into the hydraulic ram via the drive pipe (e) 2) When the pulse valve (a) snaps shut it causes a pressure wave that forces the delivery valve (b) open. Water then enters the delivery chamber (c) and leaves through the supply pipe (d).
3) When the pressure in the supply chamber falls below that in the delivery chamber, the delivery valve snaps shut and a pressure wave is transmitted back through the supply chamber, opening the pulse valve and restarting the cycle.

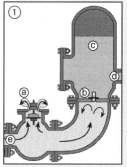
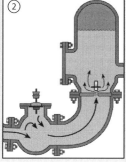
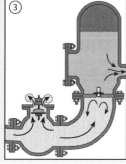

Working fall (metres)	Vertical height to which water is raised above the hydraulic ram (metres)											
	5	7.5	10	15	20	30	40	50	60	80	100	125
1	144	77	65	33	29	19.5	12.5					
1.5		135	96.5	70	54	36	19	15				
2		220	156	105	79	53	33	25	19.5	12.5		
2.5		280	200	125	100	66	40.5	32.5	24	15.5	12	
3			260	180	130	87	65	51	40	27	17.5	12
3.5				215	150	100	75	60	46	31.5	20	14
4				255	173	115	86	69	53	36	23	16
5				310	236	155	118	94	71.5	50	36	23
6					282	185	140	112	93.5	64.5	47.5	34.5
7						216	163	130	109	82	60	48
8							187	149	125	94	69	55
9							212	168	140	105	84	62
10	Litres pumped in 24 hours per litre/min of drive water						245	187	156	117	93	69
12							295	225	187	140	113	83
14								265	218	167	132	97
16									250	187	150	110
18									280	210	169	124
20										237	188	140

Table 5.4. *The output of a hydraulic ram is dependent on the fall in the drive water and the height to which you wish to raise the water.*

Box 5.4. Case study – remote holiday cottages using a hydraulic ram pump

Conversion of some barns into holiday cottages on a remote welsh hillside required the installation of a water supply to serve up to 10 people. A stream runs past above the house, about 100 metres away, but the water would have required pumping to boost pressure through the treatment system. There were additional concerns over land ownership, the presence of a bridleway and access agreements. There is a larger stream about 10 vertical metres below the house, with no access or ownership concerns. A 1.5 inch diameter hydraulic ram with a drive head of 5 metres pumps water up to 10 metres above the house where there is a slow sand filter and storage tank. Water then flows by gravity to the holiday cottages, which have a UV filter to kill micro-organisms.

Human powered pumps

Hand powered piston pumps have been used for generations, with the delivery rate varying according to the depth of the well/borehole. They are commonly specified for developing nation water supply projects but there is no reason why you shouldn't use one for your private water supply. You should ensure that you get one that is really designed for use (for example those used by development agencies, figure 5.11), rather than those designed to look nice in ornamental gardens.

It is important to specify the hand pump to the depth of the well; as discussed in the box on how pumps work, a hand pump that is relying on suction can only be used in shallow wells (where the water level is less than 7 metres below the ground surface). Hand pumps for deeper wells and boreholes must have a slightly different configuration, so that the action is more that of a submersible pump than a suction pump. Water drawn by hand pumps should be fed directly to storage tanks to avoid contamination, if these tanks are significantly above the top of the well/borehole a 'force pump' will be required (a modified type of piston pump).

Figure 5.12. A bilge pump has a rubber diaphragm in it that is moved up and down with the handle. This alternately creates suction and positive pressure, which drives water through the pump owing to the valves at either end which prevent the water flowing in the wrong direction.

Figure 5.11. A robust piston driven hand pump (Nira AF-85).

Submersible pumps	Suction pumps
Advantages	**Advantages**
Do not require priming	Simple installation
More efficient	
Can pump from greater depth	
Disadvantages	**Disadvantages**
May be difficult to remove for maintenance	Will not start automatically if priming water is lost
Installation more complex	Limited head – practical maximum 8 metres. Many models are inefficient beyond a 3 metre head

Table 5.5. A comparison of submersible pumps and suction pumps.

Hand-operated diaphragm pumps (also known as bilge pumps, figure 5.12) are very simple, and can pump to heads of around 3 metres and 40 litres/minute; it may be worth having one of these pumps for use in an emergency if your main pump fails. They are considerably cheaper than piston hand pumps (although since they are suction pumps they are not suited to deep well or borehole applications).

Electrical pumps

Advantages and disadvantages of electrical submersible and suction pumps are described in table 5.5, and their appropriateness for various situations is described below.

Most electrical pumps run from a mains electricity supply (AC current). If you wish to power your pump from a photovoltaic panel using DC current directly, you will need a pump specifically designed for the job, since the voltage produced by the panel will vary and this would tend to damage conventional pumps. Accurate site data on the solar resource available is also necessary, and the interested reader is referred to *Choosing Solar Electricity* for further information.

Pumps will generally come supplied with some kind of strainer over the inlet to prevent solids entering the pump and damaging it. Strainers with floats attached to them are also available (figure 6.7 on page 146), or you can make your own. These must be regularly cleaned so that efficiency is maintained and the pump doesn't get damaged.

Pump controls

Electrical pumps for domestic water supplies can be turned on and off by switches designed to sense changes in water *level*, or in water *pressure*. Changes in water level are sensed by ball valves or float switches. Float switches contain a lever operating a switch and a metal ball. When the ball falls it hits the lever and switches the pump on or off (depending on which way round the switch is wired). Pressure switches contain a material that changes its electrical resistance according to changes in pressure, although if the pressure switch also incorporates a pressure gauge, it may be an older design that contains a moving diaphragm (similar to that in a traditional barometer).

Protecting the pump from dry running: domestic water pumps rely on water flowing through them to keep the motor cool, so it is important that they don't operate when there is no water. With submersible pumps this is usually ensured via a float switch. With suction pumps the supply pipe usually includes a foot valve (see section on foot valves on page 129) to prevent 'priming' water from draining out.

Pump operation according to water use: this is usually achieved by sensing changes in level or pressure in different parts of the system. Float switches located in the water storage tank can be used to switch a pump on when the level falls. Very often, two float switches are used in the storage tank, one that senses low level (and turns the pump on), and one that senses when the tank is full (and turns the pump off). In other systems, the drop in water pressure in the distribution system (that occurs when an appliance is in use) is used to activate the pump via a pressure switch.

Improving pump efficiency: When electrical pumps start up they use more energy than they do during normal operation (up to 3 or 4 times as much, depending on the pump type), and they will also wear out faster. In a private water supply, the pump might therefore turn on every time someone uses even a very small amount of water (for example for hand washing), which would be inefficient and potentially noisy. It can also cause a phenomenon known as

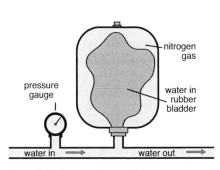

Figure 5.13. Pressure storage tank.

pump 'hunting'; where the pump continually turns on and off in response to small pressure/level changes. In order to combat these issues, pumps often incorporate pressure vessels (figure 5.13), similar to expansion vessels in heating systems. Water is pumped into a butyl rubber bladder inside the tank. The rest of the tank contains nitrogen gas. Water entering the bladder causes the bladder to expand, compressing the nitrogen, and the pressure inside the tank rises. When it hits a predetermined pressure level, the pump switches off via a pressure control switch. When a water using appliance is turned on, water leaves the bladder, the nitrogen expands and the pressure in the tank falls. When the pressure falls to a predetermined level, the pump switches on again and the tank refills. The pressure vessel is sized to allow a predetermined volume of water to be used without the pump being activated.

An alternative way of preventing the pump turning on and off frequently is to use an electronic pump controller. These consist of a pressure switch and a memory chip that is programmed (for example to pump for at least 30 seconds, or not to pump at certain times of day).

Pump choice according to water source
Groundwater sources
If you are having a well or borehole dug or drilled, the contractor will specify and install a pump for you. Submersible borehole pumps as shown in figure 5.14 are specifically designed for the purpose, and can be specified to pump to heads of over 100m if necessary. They are generally centrifugal pumps and may consist of several 'stages' if specified for deeper boreholes. If you have a shallow well or borehole (less than about 5 metres) you could in theory use a suction pump (figure 5.15), although since they are less efficient, this is not usually recommended.

Surface water sources
The type of pump best suited to surface water sources depends on the inlet structure you are using to abstract water. Submersible pumps should not be placed directly into flowing watercourses where they are likely to get damaged by silt and debris, but if the inlet structure includes an infiltration gallery, or consists of a well or sump dug next to the river bank, then a submersible pump is possible. If you are using a very simple inlet structure then you will need either to gravity feed the water to a tank that can contain a submersible pump, or use a suction pump located on the bank in a pump house.

Figure 5.14 (left). Submersible borehole pump
(courtesy of Amos pumps).
Figure 5.15 (above). Suction pump (Aspira) with pressure
switch (courtesy of the Green Shop).

The type of pump used also depends on the type of treatment system you are using (which as you will see in Chapters Three and Four, depends on the contaminants in the source water). Generally, if you need to generate water pressure for the treatment system (for example to pass the water through a filter), you will need to position a submersible pump so that it is pushing water through the treatment system, or pump to a height above it so that water can flow down through the treatment process by gravity.

Examples of system configuration
Example 1 – water from a borehole
Description: The borehole contains a submersible pump. This pumps water to a storage tank, which then gravity feeds the water to the house via a treatment system such as a string filter and a UV light.
Controls: There is a float switch on the submersible pump to prevent it from dry-running in the event of the level in the well/borehole being too low. The level of water in the storage tank (which could be the header tank in the house) is controlled by a ball valve. When water is used and the ball valve drops, the pressure in the pipe upstream of the storage tank drops, which makes the submersible pump turn on via a pressure switch. An alternative control is to have a float switch in the storage tank that signals to the borehole pump when the water level falls. Note that whilst water would continue to

flow during a power cut until the storage tank was empty, there would be no water treatment from the UV.

Example 2 – water from a borehole

Description: The borehole contains a submersible pump and there is no storage tank.

Controls: The pump is activated by a drop in water pressure (when an appliance is turned on), and pumps water directly through the treatment system and to the appliances. This type of system will generally incorporate a pressure vessel to improve pump performance. The amount of water will be limited in the event of a power cut (to the small amount available via the pressure vessel).

Example 3 - surface water supply with suction pump

Description: Water is taken from a stream which has been dammed to provide a small reservoir. Located on the bank is a suction pump that removes water from the reservoir and pumps it to a storage tank, which then gravity feeds the water to the house via a sand filter and UV.

Controls: A foot valve on the end of the suction line prevents water draining out of the inlet to the pump. A strainer is incorporated into the suction line to prevent damage to the pump. Float switches in the storage tank switch the pump on and off according to the water demand. This can be either a single float switch, or two switches, one which controls the maximum water level (in other words turns the pump off), and another to control the minimum water level (in other words, is used to turn the pump on).

Example 4 – surface water supply with submersible pump

Description: An infiltration gallery next to the river is used to access the water source. The water from the infiltration gallery flows by gravity into a water tank which contains a submersible pump. This pumps the water to the house via a treatment system.

Controls: The pump is prevented from dry-running by a float switch in the water tank. A pressure switch switches the pump on and off according to water demand.

Example 5 – surface water supply using hydraulic ram

Description: The house is located above a fast flowing stream. A pipe in the stream supplies water to a hydraulic ram pump, which pumps a proportion of the water up to a storage tank which then gravity feeds to the house via a treatment system. The remainder of the water is returned to the stream.

Controls: The hydraulic ram operates continuously. An overflow pipe from the storage tank discharges into the stream.

Other components

Most water supply systems will include other components, the most common of which are as follows:

Stop taps: Once you have your system installed there will inevitably be occasions when you want to stop the flow of water, so you will need to incorporate some valves (also known as appliance, isolation or control valves). You should try to have one between each component of your collection and treatment system, so that any component can be isolated for cleaning and maintenance. If pipes are buried this can be fiddly, as it requires manholes (and insulation). Like all connections, they add resistance to flow, and you may need to take this into account in your head loss calculations (as discussed in the box on calculating head loss on page 115). At the very least, you should make sure that there is a convenient tap near the house so you can turn off the water supply in an emergency; you don't want to have to walk halfway up a mountainside if water is pouring through your house from a header tank in the loft.

Non-return valves: These valves allow water to flow in just one direction. They can be used as a backflow prevention measure under the Water Supply (Water Fittings) Regulations and are used in situations where there is a risk of one water source contaminating another (for example on outside taps).

Float valves: a float attached to a lever sits on top of the water. When the water level falls, the lever falls and opens a valve that allows water to flow. They are the type of valve used in header tanks and toilet cisterns, and are also used in a variety of storage tanks and water treatment systems. They are used when an airgap is required to prevent contamination of a water source (and provide more protection against backflow than a non-return valve).

Foot valves: These are a specific type of non-return valve used at the bottom of supply pipes to suction pumps. They maintain a column of water in the water pipe that supplies the pump, so that it will not need priming prior to use.

Level sensors: If you want to know how much water is in your storage tank without looking in it you can put a water level probe in the tank. These are available from electronic component suppliers.

Air release valves: Air bubbles in the pipe will tend to join together and so may form air locks that can reduce flow or completely block a pipe. These are particularly likely to occur when a system is first commissioned and will usually clear if sufficient water is run through the system; you might need a pump at this stage to get the water flowing properly, even in a gravity system. If the problem persists you should check pipe joints (which can leak air into a pipe as well as leaking water out) or install an air release valve; both manual and automatic versions are available. Air release valves can easily become blocked open with suspended solids in the water, so must be checked regularly. Because the air bubbles tend to work their way to the highest point in the system, the problem can sometimes be reduced by the addition of non return valves in pipes where water is flowing downhill.

Water meters: If you are interested in working out how much water you use you can buy an inline water meter from a plumber's merchant for around £20-£30.

Strainers: Vital to prevent solids damaging pumps, strainers are discussed on page 52.

Contingency planning

Given that you won't want to be without water for more than a couple of days, it is good to have some idea of what actions to take if your water supply fails. This might be a quality failure (brought to your attention by water sampling, or following a change in taste or odour), or a total absence of water. The seriousness of this will depend on the scale of the system; if it is supplying a commercial operation or members of the public this is obviously not something that can be tolerated for long and a detailed contingency plan will be needed. Consider some of the following questions.

Quality failures: *Can the water still be drunk in the short term?* The nature of your treatment system and type of failure will determine this; if you use a 'multiple barrier' approach (see page 63) that includes source control measures and several treatment stages, then failure of one component, whilst rendering the water unsafe, may not represent a major risk for water consumed over a period of a few days (for example whilst a new part is ordered and installed). This could well be the case for failures involving dissolved parameters, but will not be true for microbiological failures.

Can the water be safely used for some household purposes despite this failure? In the case of microbiological water failures, the water will generally still be useable for toilet flushing. For some people washing and bathing in microbiologically unsafe water may be acceptable, but this is not recommended for long periods of time, or for children, owing to the risks of ingestion or inhalation.

Absence of water (freezing and drought): *Which parts of the supply are most at risk from freezing?* Even a relatively well insulated system will have areas that are more vulnerable to freezing than others. In some instances it is possible to thaw these areas (with hot water, or very careful use of a blow torch) and get the water flowing again. Be aware that pipe bursts and splits will only become apparent when the water starts flowing again; the entire system should be inspected once the thaw arrives.

How many days of storage are there once the supply has dried up in a period of dry weather? How much rain is needed before the supply starts flowing again? Because the inlet for a spring box is near the base, it can be quite difficult to tell what the recharge rate is and how much it varies. This is a worrying scenario; you may be completely unaware of the recharge rate and not have any warning of imminent supply failure.

Measuring the level in the spring box first thing in the morning (in other words after a long period without any water having been used) is a good idea, and is easiest if you fix a measuring tape to the internal wall of the spring box to monitor this over a period of time. The first indication that your use is exceeding the recharge rate will be that the overflow pipe stops flowing.

With a surface water supply it is generally much easier to predict when it is likely to dry up. It is a good idea to be familiar with what water levels and volumes are in different parts of the system under a range of conditions, so that you have an idea of when and if the supply could run dry.

Water supplies respond with very different time delays to weather conditions, with deeper groundwater supplies seeming totally independent and shallower groundwater and surface water being much more variable. The duration of water failure and how much inconvenience it causes you will dictate your actions; you might be happy to rough it for a week, but if the failure is going to be prolonged you might choose to get a tanker delivery of water from elsewhere.

All failures: *How long can I be without water?* If the water supply is particularly critical (for example if it serves multiple dwellings, or is part of a commercial operation such as a campsite or caravan park), detailed contingency plans should be made. These should include: where alternative water supplies are going to be brought in from; how the new water is going to be supplied (for example via the existing supply system, or via a bowser and water carriers); whether the water requires different or additional treatment; informing users of the situation; considering changes to fire precautions.

What caused this to arise and how can I stop this from happening again? This might be a case of being aware of imminent insufficiency, allowing you to implement whatever water efficiency measures you can in an attempt to stop the supply running dry completely. You may also consider increasing storage volumes, or being able to access an additional temporary source, or additional insulation for vulnerable points in the system.

What equipment and spare parts should I have to prepare for this in the future? This might include a store of bottled water and/or a jug filter, water carriers/ jerry cans, a pump and hoses to move water from one point to another. You may keep spare parts for your treatment system (for example a spare bulb and sleeve for UV and a spare particle filter for under sink units). You might also want to have the necessary kit for dealing with frozen pipes, a temporary water filter, and some disinfectant ready for commissioning your supply again once you've fixed it.

What are the water quality issues associated with any temporary supply? If you decide on a temporary source of different quality you will need to know what purposes this can be put to without treatment, and what treatment you would need to ensure it is of potable quality (or whether it is easier to have a store of potable water in bottles).

You may choose to invest in additional temporary storage tanks (for example IBCs) that can be filled from a tanker delivery or an alternative supply. Consider the logistics of this and discuss it with the likely supplier before it is necessary; from an access perspective these tanks may need to be situated close to a road, which may mean that they aren't plumbed in to your normal gravity fed system.

You may want them on a raised platform and to think about appropriate plumbing fittings so that you can attach a hose or pump, or put jerry cans underneath the outlet. Informal arrangements with neighbours on mains water supplies are often a good idea; a water tank on a trailer towed by a 4 x 4, filled from a hose, can be brought to site easily and repeated as necessary.

A common situation is to have a gravity fed water supply from a spring box feeding the property, but for there to be surface water at a level somewhat lower than the house. In this instance, it might be appropriate to have spare storage tanks adjacent to the house, and to keep a submersible pump and hoses so that you can rig up a temporary system to fill these tanks from the surface water if needed.

In all cases following a supply failure, bear in mind that you may need to disinfect the system once your primary water supply restarts. Your plumbing system may have been temporarily exposed to water of a lower quality, or low water conditions may have disturbed settled solids in tanks. Extra care should be taken if you use a surface water supply; a build up of sheep poo on a hillside over weeks of dry weather that gets washed into streams after one intense downpour can cause a serious spike in levels of pathogens. Additional treatment, or drinking bottled water for the first week after a supply has resumed, might be a sensible precaution in this instance. If the water doesn't start flowing despite the upstream part of the system refilling, check the system for air locks.

Short term failure or long term problem?

If your water supply runs out on a regular basis, or for unacceptably long durations of time, the basic options are to increase storage volumes, decrease water use, or find alternative supplies. Increasing storage volumes is generally the easiest option if the supply runs out for days or weeks rather than months. Increasing storage volumes is also the best option if you are able to implement more stringent water efficiency measures during times of water shortage; many people will choose to flush the toilet less, bathe/shower at work, go to a launderette, and as a result will decrease their water use at home to less than 20 litres a day.

The disadvantage with increased storage volumes is the risk that water quality deteriorates over time, so it is often wise only to bring the increased storage volume into use for fixed periods of time. You should also consider whether or not this additional storage volume is to be regarded as drinking water quality or not. This will generally be determined by where it is positioned in relation to the treatment system, and whether or not it requires additional pumping.

Implementing alternative supplies: These may be additional to the failing supply, or a complete substitute. The decision making process is similar to that for finding your initial water supply (as discussed in Chapter Two), although the severity of the water deficit will determine what lengths are worth going to. Private water supplies that are prone to intermittency are those that are highly dependent on frequent rainfall; this makes it unlikely that rainwater will be a suitable supplementary supply, although it is often the first considered as it is the easiest.

For supplies where there is a significant time delay for the water supply to return after rainfall events (for example for the water to permeate the soil and saturate it sufficiently for the ground to become water bearing, or the surface water supply to return) capturing rainwater from the roof may be useful; a single summer thunderstorm may fill a storage tank fed by the roof of the house, but have no discernible impact on a groundwater supply.

Final touches – commissioning, inspection and record keeping

Commissioning: Once you have assembled your water supply the components will need to be flushed out and disinfected before use, to kill any pathogens that may be present; this may also be necessary following maintenance. The final concentration in the structure you are disinfecting should be 0.2% chlorine by mass.

Liquid bleach (sodium hypochlorite) contains about 5% free chlorine, so to make a 0.2% solution by diluting it you will need to add 1 litre of bleach to 24 litres of water. If the structure you are disinfecting already contains water (for example a borehole), you will need to work out how strong the solution should be to make a final concentration of 0.2% when you have added it to the existing volume in the borehole.

If you can drain the structure (for example a spring box or a storage tank), it may be more appropriate to do this, and then scrub the walls with the solution, rather than attempt to fill the whole structure. The chlorine should remain in the structure for at least 12 hours, after which water should be run to waste until it no longer smells of chlorine. Suppliers of swimming pool chemicals can provide chlorine in tablet form (calcium hypochlorite), which may be more convenient. Full details of recommended commissioning procedures are given in BS6700.

Inspection: No private water supply will be 'fit and forget'. You should develop an inspection and maintenance regime based on table 3.3 on pages 65-68. Regular inspections of the various parts of the system will be a minimum; other elements will need replacing, servicing, or some other maintenance procedure in order to maintain good water quality. Inlet devices should be checked to see that they are not blocking (surface water supplies) and that source protection measures such as fences, lids and so on (ground water supplies) are in place (as discussed in Chapter Two). Storage and header tanks should also be inspected periodically to make sure that water quality doesn't deteriorate during storage. It is a good idea to have all of this information collated in one place as a management plan.

Record keeping: It is highly advisable to keep records of what your supply consists of, what work is done on it, inspections, and test results relating to it. The checklist in Chapter Three is a good starting point for what type of information is worth having to hand. These records will be useful if you need any work done on the system, if the local authority visits it, on change of ownership of the property, but most importantly if anything goes wrong.

Summary

Your water supply system will include some kind of inlet structure, storage, pipe-work, control mechanisms, cleaning processes and possibly an energy source. The most appropriate choice for each element is affected by a number of factors, including the type of source, how clean it is, the pressure required, and the relative heights of the source and the house. The latter will need to be calculated, and sources of head loss accounted for to work out whether or not a pump is required. Renewably powered pumps are available but their use is often site specific. Planning ahead will allow you to reduce both the likelihood and severity of water shortages.

Further reading

BS6700. Design, installation, testing and maintenance of services supplying water for domestic use within buildings and their curtilages – guidance on pipe diameter and head loss calculations, commissioning tests and so on.

Choosing Solar Electricity, Brian Goss, CAT Publications, 2010. Available from CAT's Eco store.

Water Regulations Guide – guidance from WRAS (Water Regulations Advisory Scheme) on the Water Supply (Water Fittings) Regulations, 1999.

Water Distribution Manual (Oxfam Water Supply Scheme for Emergencies). Considers the physics of pipeline design.

Water Pumping Manual (Oxfam Water Supply Scheme for Emergencies). Considers pump selection, installation and maintenance.

Water Storage Manual (Oxfam Water Supply Scheme for Emergencies). Oxfam guides are downloadable from www.oxfam.org.uk

Water Supply (Water Fittings) Regulations, 1999 – regulations that specify types of valves required according to contamination risk. Downloadable from www.hmso.gov.uk

Windpumps, Barlow, ITDG Publications, 1993. Estimating the wind resource and specifying wind pumps for water supply. Available from CAT's Eco store.

Chapter Six

Rainwater harvesting

What to use rainwater for – how to use it – parts of the system – doing the calculations – maintenance – environmental issues - economics

In most parts of the UK, the water that falls on the roof of your house is collected in gutters, directed into down pipes and then gets discharged into the drains as if it is a waste product. This causes a range of problems (discussed on page 35) and you could look upon this rainwater as a resource that you can collect and use for your own purposes. After all, it's clean, free and relatively straightforward.

This chapter describes how and why you might do this. Much of the detail on storing, moving and cleaning water that has been discussed in Chapters Two to Five in the context of private groundwater or surface water supplies applies equally to rainwater, but is repeated here for the benefit of those readers with mains water supplies, so this chapter may be read in isolation.

What can rainwater be used for?

How you use your rainwater depends on how much you are prepared to store and how much trouble and expense you are willing to go to, to clean it. The simplest measure is to install a water butt at the bottom of each rainwater downpipe and collect untreated rainwater for use in the garden. This very basic level of rainwater harvesting is probably as far as most people should go, and is discussed in detail in Chapter Eight.

If you collect more rainwater than you can use in the garden the next priority is to use it for non-potable purposes within the home, so that you can reuse it without concerning yourself too much with cleaning. Referring to figure 1.7 on page 19, the major non-potable water demand in the home is the toilet, so this would be the most obvious appliance to use rainwater for.

It may also be appropriate to use your rainwater supply to run your washing machine, particularly if your mains water supply of water is very hard. In practice, if you are using rainwater for both toilet flushing and garden use (the most common scenario), you may be able to meet most of these combined demands in the winter, but a much smaller proportion in the summer.

If you are frugal with your water use, it can be possible to meet all your domestic water needs using rainwater from your roof, but given the infrastructure required, it is rarely justified in the UK. This is in contrast to many other countries; isolated properties in Australia routinely rely on rainwater for all their domestic needs, largely due to the lack of other water sources and the absence of mains water. If you do not have a mains water supply, consider the water sources discussed in Chapter Two before investigating rainwater harvesting. A UK example of a water-autonomous house is described in the case study below.

Box 6.1. Case study – the self-sufficient house

This town house was built according to the specifications of eco-architects Robert and Brenda Vale and is located in Southwell, Nottinghamshire, where the annual rainfall is 0.6 metres a year. The total roof area collected from is 140m² and is made of tile and glass. Water from the roof passes through a coarse filter to storage tanks in the basement that have a capacity of 28,500 litres (28.5m³). The water from these tanks is pumped into a sand filter and from there into an additional storage tank.

From this tank the water is pumped to a 250 litre header tank in the loft, which feeds appliances. A spur from this pipe pumps water through activated carbon filters to supply the drinking water needs of the house. Given the limited water supply, a compost toilet is used instead of a flush toilet. The system has been working for ten years and there has never been a shortage of water, in part due to the exceptionally low water consumption of the residents (around 35 litres per person per day).

How to use rainwater

Good practice for rainwater harvesting systems is covered in British Standard BS8515. Decisions regarding RWH systems generally centre around two issues; the system configuration (discussed below), and the storage volume (discussed in box 6.2 on page 150).

System configurations for a rainwater harvesting system

A rainwater harvesting system generally has the following components:

- Collection surfaces and gutters.
- Filter.
- Storage.
- Pump.

• Pipe-work.
• Cistern/header tank.
• Control system.

There are three basic layouts for these components, illustrated in figures 6.1, 6.2 and 6.3. Directly pumped systems are the most common, but have the disadvantage that there will be no water during power cuts, or if the pump fails. Indirectly pumped systems are usually only retrofitted to properties where the toilets are fed from an existing header tank in the loft. Gravity fed systems seem simple but commercial versions are not widely available and installations are often DIY systems assembled from the various necessary components. The relative merits of these system configurations are discussed in table 6.1.

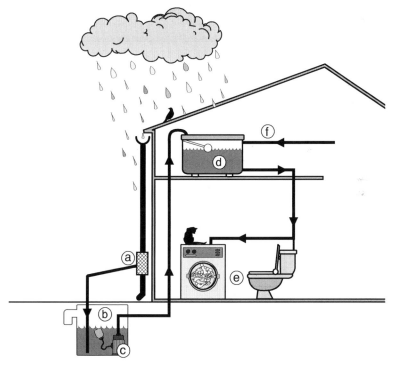

Figure 6.1. Indirectly pumped rainwater harvesting system. Rainwater passes through a filter (a) and into the storage tank (b). Water is pumped (c) into a header tank (d), from where it flows to the appliances (e). Mains water backup (f) is provided to the header tank via a suitable air-gap.

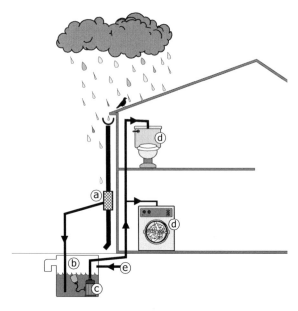

Figure 6.2. Directly pumped rainwater system. Water passes through the filter (a) and into the storage tank (b). It is then pumped (c) directly to the appliances (d). Mains water backup (e) is to the storage tank via a suitable air-gap.

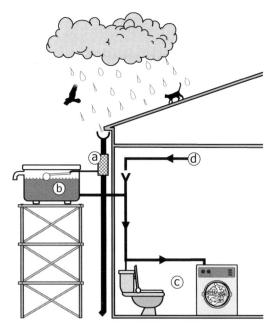

Figure 6.3. Gravity fed rainwater system. Water passes through the filter (a) and directly into a roof-level storage tank (b), from where it flows into the appliances (c). Mains water backup (d) is via a tundish (see figure 6.8) or other suitable air gap.

	Indirectly pumped (figure 6.1)	Directly pumped (figure 6.2)	Gravity fed (figure 6.3)
Advantages	Generally simplest system to integrate into existing buildings. Can be quieter than directly pumped systems and more energy efficient, as pump doesn't operate so frequently.	High pressure at appliances so rapid fill rate. Avoids need for separate header tank.	No energy requirement. Simple and low cost.
Disadvantages	Low water pressure at appliances (usually fine for WCs, but may cause problems with washing machines).	No water to appliances during power cuts or if pump fails. Mains backup to underground tank incurs additional energy cost for pumping.	Low water pressure at appliances. Unlikely to be able to feed upstairs WCs. Weight bearing capacity of structure must be assessed. Decline of water quality (if indoors), risk of freezing (if outdoors).

Table 6.1. Relative advantages and disadvantages of the various configurations of RWH systems.

Component details

Where to collect rainwater from

When we consider the surfaces we can collect rainwater from, we are concerned with the effects that they have on water *quantity* and water *quality*. The most obvious place to collect rainwater from is the roof of your house, but if you have other substantial roof areas close to the house, collecting from those might also be appropriate. In single-household schemes collecting rainwater from the ground is rarely appropriate, since it will require additional cleaning unless the water is just used for garden watering.

In order to estimate how much rainwater you can collect from your roof, you need to measure its area. This is the amount of ground that the roof covers, not the surface area of the roofing material itself (which will be larger if you have a sloping roof, see figure 6.4).

Figure 6.4. The amount of rainwater you can harvest is determined by the ground area covered by the roof, not its surface area.

You should try to avoid collecting rainwater from roofs that have trees overhanging them; either remove the trees or don't collect the water; it makes more sense to remove the contaminants at source than to try and clean the water once it has already become dirty. In addition to leaf fall, bear in mind that bird droppings are exceptionally high in micro-organisms (a study of seagull droppings found 2.6 million faecal coliforms per gram of droppings, in addition to other pathogenic bacteria such as Campylobacter), so you may also want to avoid feeding birds in the garden.

The slope and material of your roof also influence the water quality. Smooth materials, such as glass, slate, glazed tiles and metal are best. Turf or sedum roofs are not recommended for rainwater harvesting systems if you are reusing the water inside the house, since the rainwater may become coloured and will collect grit and silt that can corrode your appliances. Rough materials, such as asbestos, felt or wood often have moss or lichen growing on them which can be a particular problem as it then provides a food source for organisms in the rainwater storage tank. Wood may also impart colour to the rainwater from tanins.

Rainwater is slightly acidic and will dissolve small amounts of lead roof flashings. Assuming you are not going to drink the water this isn't a problem, although lead gutters and downpipes should be avoided as they may result in lead levels being too high for rainwater to be used on edible crops in the garden.

In schemes where demand is high and large paved surfaces form part of the development, collection from ground surfaces may be appropriate. Additional treatment beyond standard rainwater filters is likely to be necessary, the nature of which will depend on the type of surface collected from and other measures in place (for example a petrol interceptor might be required if the collection area included significant car parking areas, gully traps to collect silt and bar screens to remove litter, leaves and so on). If the scale of the system is even larger, collecting from several hectares of ground, then the system should be designed in the same way as the surface water systems discussed in Chapter Two.

Your roof type will also affect the yield coefficient; the proportion of rainwater landing on the roof that is actually available for use, compared to that which evaporates. The steeper and smoother the roof surface, the better the yield coefficient will be. Flat roofs will lose as much as 50% of water to evaporation, particularly during brief rainfall events. The yield coefficient is also influenced by the pattern of rainfall events; less water will be lost to evaporation during sudden showers than the same amount of rain falling as a long period of drizzle.

Gutters and down pipes – moving the water from the roof to storage
Gutters should have a 1½-2% fall wherever possible, and there should be sufficient brackets to prevent any sag, since standing water will tend to become contaminated by any solids that collect in the gutters. As with roofing materials, smooth surfaces are preferable. All rainwater goods should be easily accessible to allow cleaning.

Some older houses have the waste water from baths and sinks entering the rainwater disposal system rather than into the foul sewer (figure 6.5), and it is essential to disconnect these (and connect them to the foul sewer) to prevent rainwater being contaminated by this much dirtier water. To collect rainwater from both sides of your roof it may be necessary to modify the layout of downpipes, but you should make sure that gutters are not overloaded. You can calculate the necessary gutter sizes and number of downpipes using the guidance in the BRE Good Building Guide 38, Disposing of Rainwater, or in Part H of Building Regulations.

Figure 6.5. If waste water pipes from showers, washbasins etc are connected into your rainwater goods, you will need to divert them into your foul drainage system before making use of your rainwater.

First flush mechanisms: The first few millimetres of rain during a rainfall event will pick up the majority of the contaminants from the roof and gutters and effectively wash the roof; later rainfall will be much cleaner. It is possible to get various types of diverting mechanisms that automatically prevent the first few litres of water from being collected into the tank, thereby improving the overall quality of the water collected. These are most appropriate in situations where the rainfall pattern is of sudden showers, rather than continual drizzle, otherwise the percentage of water collected is considerably reduced (on average the devices reject the first 40 litres or so of water per rainfall event). They are widely used in both Australia and the USA, but are redundant when good quality filters are used and are rarely used in the UK.

Cleaning the water
Whilst it's in the sky, rainwater is very clean. It may contain a few dissolved substances, including air-borne chemical pollutants, but when it hits your roof or the ground, it immediately has the potential to become contaminated by leaves and moss, dust/grit, and pathogens from birds and small mammals. If collected at ground level, there is the potential for contamination by hydrocarbons such as oil, petrol and diesel, and a considerably increased load of dirt, silt and animal faeces.

With suitable roof and gutter materials and good housekeeping, simple filtration provides sufficient cleaning for the water to be used directly for toilet flushing without any disinfection. Filters can be positioned in the downpipe or underground, and which is the best option will depend upon how many downpipes water is being collected from. A typical downpipe filter is shown in figure 6.6. Collection efficiency for these can be around 95% (when clean), but will fall to below 30% if the filter isn't cleaned regularly (algal growth blocks the holes in the filter mesh).

Ease of access for cleaning should therefore be considered when choosing a filter. Filters that include a mesh that prevent water flowing through the downpipe if the filter itself becomes blocked should be avoided; these are liable to cause overflowing gutters and dampness in walls. If rainwater is to be used for anything other than the toilet and washing machine, then the water quality recommendations in table 3.1 on page 62 should be referred to.

Storage
Even if the incoming water is clean, contamination problems can be caused

Figure 6.6. A Wisy downpipe filter. These filters replace a section of your rainwater downpipe and divert the water into a storage tank. When water flows down a vertical pipe, it tends to stick to the edges of the pipe by surface tension. In the filter, a section of downpipe is removed and replaced by a mesh, as shown in the cutaway section. Water passes through the mesh and through the outlet to the rainwater tank.

by poor storage tank design; frogs, insects, rodents and worms may all get into your tank if it isn't properly sealed at the inlet, outlet, lid and overflows. Any pathogens in the water have the potential to breed, particularly if there is any organic material contamination to provide a food source. Storage tanks for rainwater should include the same design features as those for potable water, as illustrated in figure 5.1 on page 111.

Tanks are best situated underground and in the dark, where possible, as the lower light and temperature levels will reduce the potential for pathogen growth. However, putting the tank underground increases the expense and can complicate the installation of drains and overflows, particularly in existing properties. A range of underground tanks are available, include plastic, fibreglass and concrete.

The best environmental option amongst these are plastic tanks that are rated such that concrete backfill is not required (backfill requires more concrete than would be necessary to build the entire tank from concrete). It is possible to have rainwater tanks in a basement, although access difficulties may limit the choice of tank, volume of storage and increase the cost.

Pumps, controls and back up water sources
The principles of how RWH systems are controlled and operated are similar to those discussed in Chapter Five in reference to other private water supplies. Following storage you will need to pump rainwater back into the house,

unless you have a gravity fed system (figure 6.3). Commercially available RWH systems will come complete with a suitable pump and controls. The way in which water enters the pump varies between systems; submersible pumps often have an inlet point built into the body of the pump that includes a filter, others may have a port for an inlet pipe. Submersible pumps will usually be suspended on a chain to allow them to be removed.

The inlet (known as the extraction point) will either be at a fixed level in the tank (positioned to ensure that settled solids from the base of the tank don't enter the pump), or it may be a floating extraction point (figure 6.7 shows a proprietary model, but an equivalent can be made out of filter cloth and a float from a ball valve). Bear in mind that many systems are very poorly designed from an energy efficiency perspective; there is around a ten fold difference in energy consumption between systems available on the UK market.

It is assumed that a) you will have an additional water supply (be it private or mains) that you can use to flush your toilets during periods of low rainfall, and b) you will want to have this back-up system acting automatically, without you having to check levels in tanks or operate valves. In most cases this back-up supply will be your mains water with a failsafe measure to prevent rainwater flowing back into your potable water system (figures 6.8 and 6.9).

In direct feed systems (figure 6.2), this mains water back up will be to your underground storage tanks. When the mains back up system causes water to enter the storage tank, it is designed to minimise the use of mains water, by only allowing small volumes of water in (so that a large tank void is still available should it start to rain). If the system is used for garden watering, you should be aware that when water company water use restrictions are in force it will be prohibited to use this rainwater tank for garden hoses (as it is effectively mains water). This limitation would not apply to indirect feed systems where the mains water back up is to the header tank in the loft.

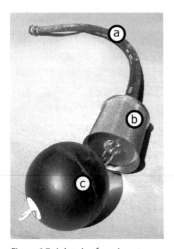

Figure 6.7. Inlet pipe for rainwater pump. Water enters the pipe (a) via the mesh filter (b). The position of the filter is kept close to the surface of the water by the float (c) which helps prevent both floating scum and settled solids from entering the pipe.

Safety and regulation

Many people are surprised by the officious nature of regulation in relation to RWH systems, given that most will be used just for toilet flushing. This is contrasted to the (lack of) regulation in many other countries where rainwater is often used for drinking. However, a significant part of the regulatory emphasis is to prevent the backflow of rainwater in to the rest of your plumbing system and back into the mains supply; this is a public health issue and could have much wider effects on people other than the users of the building with the RWH system. Also, given that most RWH systems in the UK are installed where mains water is available, it is not unreasonable for the benchmark comparison to reflect the impressive reliability, safety and low cost of mains water.

Labelling: Both BS8515 and WRAS have recommendations on labelling of pipework and appliances connected to rainwater systems. In essence it requires all pipes containing reclaimed water to be labelled in a specific way, to prevent inadvertent connection of reclaimed water pipes with potable supply pipes. Pipes carrying rainwater should never be blue (as this is the usual colour for potable water pipes), but instead should be green, or green and black. Insulated pipes need labelling outside the insulation layer as well as on the pipe itself. Labelling of reclaimed water supply points (taps etc) is also required to indicate that the water is not suitable for drinking.

Backflow: Under the Water Supply (Water Fittings) Regulations, 1999, varying degrees of precaution must be taken against backflow, depending on the degree of contamination of the water. Contaminated water is therefore divided into a number of fluid categories. Rainwater falls into Fluid Category 5 (since it may constitute a serious health hazard owing to the likelihood that it contains pathogens). The regulations therefore require the fitting of a 'Type AA, AB, AD or AUK1 air-gap', and this will be incorporated into the tank or the pipework (as shown in figures 6.8 and 6.9).

Informing the water supplier: Under the Water Supply (Water Fittings) Regulations, you must notify your water supplier prior to commencing work and send them details of the proposed installation; this is so that they can check that requirements relating to backflow have been complied with. A knock on effect of this is that you should therefore expect these details to be passed on to the billing department, which may well result in your sewerage charge being decoupled from your mains water charge, as discussed on page 14 of Chapter One.

Water quality: BS8515 recommends that water quality from RWH

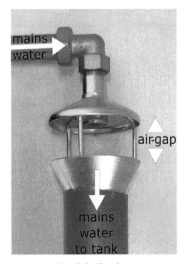

Figure 6.8. Tundish. This device is one way of providing the necessary air-gap to prevent backflow of rainwater into your drinking water supply. Mains water falls through the air-gap into the collection pipe below, and from there to a storage tank. If rainwater were to fill the collection pipe, it would overflow from the tundish rather than enter the mains water supply.

systems for toilet flushing should not exceed 1000 coliforms/100mls (see table 3.1 on page 62 for details of water quality recommendations for other purposes). In practice, the safety concern with using rainwater for toilet flushing is not related to the potential dangers of water when it is in the toilet cistern or bowl (an already pathogen laden environment), rather it is to do with the risk of cross-contamination of other parts of the water supply system. Some installations incorporate UV disinfection, although since UV is only effective on water with no solid content or colour it is of limited usefulness for rainwater that has only passed through a coarse filter.

If you are on a private water supply and have no mains water connection, you are not obliged to comply with these regulations since there is no risk of contamination of the public mains water supply. However, they do represent best practice and given the potential health consequences of cross-contamination we would strongly advise you to follow them.

Sizing the storage tank

Rainfall is intermittent, whereas your daily demand for toilet flushing is likely to remain fairly constant, and demand for garden water will coincide with the lowest rainfall times. This mismatch means that in most instances it isn't practical to meet the entirety of these demands. If you are purchasing a rainwater harvesting system, the vendor will calculate your tank volume for you, but you may wish to check that their recommendations are in line with standards.

Early UK systems usually had tanks that were rarely full and led to the water quality deteriorating as the water stagnated. The emphasis has shifted towards smaller tanks that overflow during prolonged rainfall events.

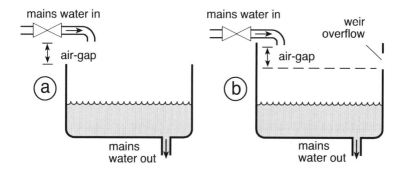

Figure 6.9. (a) Type AA air-gap with an unrestricted discharge. (b) Type AB air-gap with a weir overflow. The air-gap must be 20mm or twice the diameter of the inlet pipe's bore diameter, whichever is greater.

Conversely, if the system is required to provide storm water control in the event of heavy rainfall the tank will need to be much larger, and a further set of calculations is needed. BS8515 includes three different methods for calculating tank volumes:

Simplified approach – for households with constant daily demand. The storage volume can be read off a graph (for example figure 6.11).

Intermediate approach – uses simple formulae to calculate supply and demand, and uses the smaller number for a suggested storage volume, as described in the box *Calculating RWH system size* in box 6.2 on page 150. Suitable for households, and for simple non domestic systems (those with fairly constant demand, such as offices).

Detailed approach – for larger systems, or where daily demand varies, or where stormwater attenuation is to be incorporated. Also recommended for systems where provision of rainwater for garden watering is an important feature. This approach requires daily rainfall data, collected over at least a three year period, and involves a model comparing supply and demand over the duration.

The two simpler methods are compared using a typical four person household in box 6.2 on page 150.

The yield of rainwater depends upon the following factors:

• Rainfall.
• Collection area.
• Nature of collection surface (determines the yield coefficient).
• Efficiency of any cleaning systems (the filter coefficient).

Material and pitch	Yield coefficient
Roof tiles at usual pitch (35-55°)	0.75-0.9
Flat roof with a smooth surface	0.5
Flat roof with gravel, turf or sedum	0.4-0.5

Table 6.2. Yield coefficients for various roof materials and pitches indicating how much of the rain falling on the roof is available for use.

Rainfall: The national average rainfall in the UK is almost exactly 1 metre per year, and the regional variations are shown in figure 6.10. Even within very small geographical areas rainfall can vary dramatically due to land topography. Annual data from around 30 stations around the UK is available on the Met Office website free of charge, but if you are close to the coast or to mountains use of more local data is recommended. The seasonality of rainfall also varies, and for larger systems or where supply is critical, daily or monthly rainfall data might be necessary.

Yield coefficient: The roof material and how sloped it is will affect how much of the water that falls on the roof is actually collected. Typical yield coefficients are given in table 6.2.

Filter coefficient: Depending on the type of filter you use and how often you clean it, you can expect filter efficiencies to be over 80%. Most are designed to maintain this efficiency over a wide range of flow rates and you should refer to the manufacturer's data for a filter coefficient.

Box 6.2. Sizing storage systems according to BS8515

Simplified approach

This method assumes a fairly constant water use over the year and a standard roof pitch. The relevant storage volume is read off a graph from BS8515; note that there are different graphs depending on the number of people in the household. Taking the correct graph, draw a straight line upwards from the x axis at the point corresponding to your measured roof area, until you meet the diagonal line that is closest to your annual rainfall (in mm). Then draw a horizontal line from this point across to the y axis. This is the recommended storage volume.

Intermediate approach

The storage capacity is calculated twice; once from the available supply, and again from the demand.

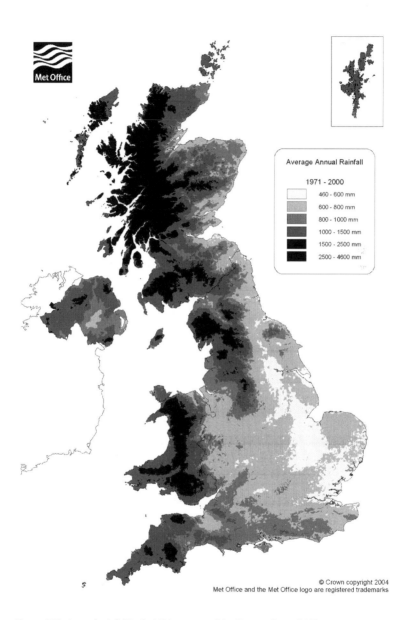

Average Annual Rainfall

1971 - 2000

460 - 600 mm
600 - 800 mm
800 - 1000 mm
1000 - 1500 mm
1500 - 2500 mm
2500 - 4600 mm

Figure 6.10. Annual rainfall in the UK (courtesy of the Meteorological Office).

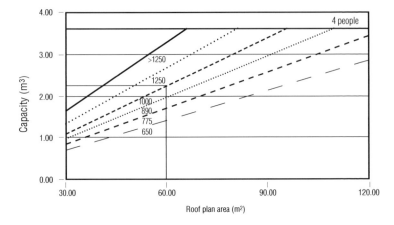

Figure 6.11. Graph showing storage tank size versus roof area.

Annual rainwater yield is calculated according to:

Annual yield (litres) = collecting area (m²) x yield coefficient x rainfall depth (mm) x filter efficiency

Annual demand is calculated according to:

Annual demand (litres) = daily requirement per person (litres) x number of people x 365

The recommended storage volume is then 5% of whichever is the smaller value; annual yield or annual demand.

Example calculation

Supply: The house has a suitable roof area of 60m². The roof is slate and has a steep pitch, the yield coefficient is therefore taken as 0.8. A filter coefficient of 0.8 is used, and annual rainfall is 1000mm.

Demand: There are 4 occupants, none of whom are at home during the day. A dual flush toilet is installed, and water use for toilet flushing is estimated at 4 flushes per person per day, with an average flush volume of 5 litres. There is no garden.

Tank size according to simplified approach: Line drawn upwards from the 60m² point on the x axis in figure 6.11 to the point where it intersects with the diagonal line corresponding to an annual rainfall of 1000mm. A horizontal line from this point to where it intersects the y axis suggests a tank volume of about 2.25m³.

Tank size according to intermediate approach: Annual rainwater yield is 60 x 0.8 x 1000 x 0.8. 5% of this yield is 1.92m³. Demand is 4 x 20 x 365. 5% of this is 1.46m³. The smaller of these two values is used for the storage volume.

Note that if, like in this example, the yield exceeds the demand, it could be worth considering using the rainwater for other purposes, such as the washing machine. In this case the recommended tank size is higher at 2.19m³ (assuming an additional 10 litres per person per day for the washing machine).

Note that the graph suggests smaller storage volumes in areas with low rainfall; this reflects the fact that tanks are rarely going to be full. The graphs used for the simplified approach are based on the equations used in the intermediate approach.

When sizing storage tanks, bear in mind that the usable storage volume (known as the working capacity) is lower than the total physical tank volume; the tank will never be fuller than the level of the overflow pipe, and additionally because the outlet point from the tank should be secured above the base the bottom 10cm or so of water will not be used.

Maintenance, monitoring and reliability issues

Many early UK RWH systems suffered from reliability issues. As yet, there has been no independent review of systems looking at maintenance issues, or comparing actual performance with that predicted, but a number of common problems exist.

In direct feed systems where mains water enters the rainwater storage tank, the rainwater element could be turned off or malfunctioning without the user knowing it. There are published incidents of owners of systems in commercial buildings believing them to be working correctly, only to discover that they have been deliberately turned off by maintenance staff. Other problems seen by the author, or reported to CAT include:

- Pumps overheating and failing (as a result of the filter on the suction hose inlet being totally blocked and imploding). Pump failure in direct feed systems results in no water being available.
- Filters becoming blocked with algal slime. Measured filter efficiency was 20%, but returned to 97% after cleaning with a scrubbing brush. Given that many downpipe filters are difficult to remove, the likelihood of suitably frequent maintenance is low.
- Washing machine failure, apparently due to blockage of water inlet pipe by fine particles and subsequent overheating. Guarantee on machine invalidated by use of rainwater.
- WCs filling slowly or not at all due to fine particles blocking inlets on the float valve.
- Mosquito larvae and decomposing animals (frogs and rats) in tanks.
- Tangling of float switch cable leading to the mains water back up running continuously and overflowing (one system discharged approximately 45,000 litres of water to the drain before it was noticed by the owner). An automatic cut off valve on the mains water back up that is activated by the overflow would have been a straightforward preventative measure in this instance.

- Faulty level sensors causing tanks not to fill.
- Discoloured water (greenish hue) from systems fed by green roofs, brownish hue from systems fed from timber roofs. Both these incidents led to users flushing the toilet prior to use, thereby drastically increasing water use, and the load on the sewage treatment system.

Whilst many of these maintenance issues are easily avoided (see table 6.3 for suggested maintenance regimes for RWH systems), it is clear that RWH systems are not suited to people who expect 'fit and forget' systems similar to mains water.

System component	Notes	Frequency
Gutters/downpipes	Check for leaks, blockages, sags	Annually
Filter	Clean	Monthly
Storage tank	Check for leaks and build up of debris	Annually
Pumps and pump control	Check joints for leakage, check pump for corrosion, pump inlet for blockage	Annually
Control unit	Check functioning for accuracy where needed (e.g. of level sensors)	Annually
Back up supply	Check functioning and that air gaps are maintained	Annually
Wiring	Visual inspection	Annually
Pipework	Check for leaks and that overflows are clear	Annually

Table 6.3. Maintenance of RWH systems should be in line with manufacturer's guidelines, or using the above as a starting point. Frequency of each element can be modified in the light of experience.

Environmental considerations surrounding RWH

Before investing in a rainwater harvesting system, you should think carefully about what your motivation is. As discussed in Chapter One, by far the largest environmental impact relating to your water use is heating it, so measures related to hot water efficiency and low carbon water heating systems should be your first concern. General water efficiency measures (as discussed in Chapter One) and collecting as much water for the garden as you can use (as discussed in Chapter Eight) are also priorities.

RWH systems are often touted as being 'environmentally friendly'. However, there is no evidence at all to support this view (on the contrary, there is a considerable amount of work to refute it). Both independent academic reviews and a UK Environment Agency report have found that the environmental costs of rainwater harvesting systems outweigh the benefits. As such they are being extensively mis-sold by many UK suppliers. There is room for system improvement (such as using more efficient pumps and using different materials for the storage tanks), but it is also worth noting how small the actual numbers are; the CO_2 impact of supplying water to your home is around 100g CO_2 per household per day. For most people this is about 1/600th of your total daily CO_2 impact, so if being environmentally friendly was your motivation for considering alternative water supplies, there will certainly be higher carbon areas of your life you should work on before thinking about RWH systems!

Water scarcity is often used as justification for RWH systems. However, as with CO_2 impact, this claim does not stand up to scrutiny. Areas with a scarce water resource are also those with low rainfall (compare figure 1.2 with figure 6.10). Consequently, at the times and places when water supplies are likely to be most stressed, rainwater systems are least likely to be able to meet any shortfall. This is reflected in the recommendation that storage volumes are smaller in low rainfall areas (as illustrated in figure 6.11).

Protection against flooding is another poorly thought out justification for RWH. There is a fundamental mismatch between requirements for RWH systems (in other words that the tanks are normally full of rainwater that can be used for toilet flushing) and stormwater attenuation (that the tanks are normally empty and therefore able to slow down runoff from the site). When planning permission is granted for new developments, a maximum rate of water run off allowed from the site will be stipulated, and this will often mean that some way of slowing down the movement of storm water off the site will be needed.

This is often provided by storage tanks, and in some instances the stored water from these can then be used for non potable uses such as toilet flushing. BS8515 discusses how to calculate the storage volume for these systems; much larger tanks are needed for storm water attenuation than is usual for RWH systems, and given that stormwater is often of very low water quality, additional filtration and treatment may also be necessary if this water is to be used. A RWH system does not decrease the amount of storage volume required for flood attenuation; the volumes of water that might be used for toilet flushing are several orders of magnitude too small.

Using RWH to meet water requirements under the Code for Sustainable Homes (CSH)

As discussed in Chapter Two, Part G of Building Regulations and the Code for Sustainable Homes (CSH) stipulate water efficiency requirements for new houses. It can be difficult to meet the water requirements for higher CSH levels using water efficiency measures alone. Consequently, rainwater and greywater systems are being installed in some of these properties.

Given how flawed the water calculator approach is, this is a source of frustration to many developers, who are faced with spending thousands of pounds extra per property to install rainwater systems that have been dogged with maintenance issues, do not provide useful water volumes, and are environmentally worse than mains water supply. This situation looks set to continue unless developers mount a legal challenge.

If a RWH system is absolutely necessary to meet CSH requirements, then ensure its environmental impact is minimised; use a polyethylene tank rather than fibreglass, and ensure it is of a class that does not need concrete backfilling of the excavation. Consider the lower energy RWH systems (that use 12V DC pumps), and systems with control panels that indicate whether the system is operating on mains water backup or is filling with rainwater, and how full the tank is. Also consider metering both the mains and rainwater in the system, so that system performance can be monitored.

Will a domestic rainwater harvesting system save me money?

A professionally installed RWH system will cost upwards of £3000. Systems are typically more difficult and more expensive to retrofit into properties than if they are installed as part of the initial build process. You should expect to replace some components (for example the pump) after 5-10 years, and do regular maintenance (as indicated in table 6.3), both of which you may need to pay a contractor to do (particularly if there is a guarantee applying to the system).

If you are on a water meter your sewage bill is calculated on the principle that 95% of your water use is returned as sewage (the discrepancy is an allocation for outdoor water use). Since installing a RWH system will decrease your use of mains water, you might also expect to see your sewage bill go down. However, you are legally obliged to tell your water supplier that you are installing the system, so in general they will not decrease your sewage charge.

In rough terms therefore, given an average UK bill (metered users) of around £330 per year (for water and sewage), and likely water savings of around 20-30%, the system might save you £30 a year, and that is before you account for any maintenance or capital costs. A comprehensive (entire PhD!) study that modelled several thousand different configurations of system demonstrated that mains water would need to cost more than £10/m^3 before RWH systems made financial sense. Regional differences in rainfall and water bills mean that systems may be economically viable in some instances (for example in the South West of England), but do your own calculations rather than relying on those from RWH system suppliers.

RWH beyond domestic scale systems

The situation in which rainwater harvesting has been used with most success in the UK is in agriculture. The combination of large collection areas, water demand being critical to operations and the relationship to pollution control and flooding can lead to short payback periods and environmental benefits in many types of agriculture, including dairy farming, intensive livestock rearing, polytunnel growing and operations where slurry management is an issue, or where the farm is in a nitrate vulnerable zone. Case studies and advice are available from ADAS. Many farmers also have the capacity to consider water management over significant land areas, and this can include measures to slow surface water runoff and collect and use this water as part of diffuse pollution control.

Non-domestic buildings with high non potable water demands may on occasion be good candidates for RWH systems. These buildings often have a large roof area, and a high proportion of their water demand is for toilet flushing (figure 1.7). These buildings are also likely to be on metered water supplies, and may also be charged for disposal of rainwater in addition to sewage. Specific industrial applications that have a high demand for process water that doesn't need to be of potable standard may also be well suited to rainwater harvesting.

There are also a number of community scale projects where rainwater is collected from a number of neighbouring houses and stored centrally, but only piped back into a proportion of them. As yet, few, if any, of these projects have been independently monitored to see if predicted water savings have been realised. There is as yet no standard approach to ownership, responsibility, maintenance and revenue issues arising from this type of communal arrangement.

Summary

It is rarely appropriate to try and meet your entire domestic water needs with rainwater in the UK. If you are going to collect rainwater, first implement water efficiency measures, with rainwater collection for the garden being the next priority. The situations in which RWH for toilet flushing is appropriate for domestic situations are few and far between; it is a technology more suited to industrial/commercial scale or as part of a catchment management strategy for pollution control in farms. If you are going to install a system, careful attention to detailing of collection surfaces, gutters, filters and storage tanks allows collected rainwater to be of good quality, and calculations of rainwater yield should be considered in light of water demand in the building.

Further reading

Approved Document H, Drainage and Waste Disposal, Building Regulations, 2000 – how to calculate roof gutter size according to rainfall and roof area.

BS8515:2009. Rainwater harvesting systems – Code of Practice.

BNWAT 19. Market Transformation Programme. Alternative sources of water – greywater and rainwater reuse: innovation briefing note. Available from MTP website.

Disposing of Rainwater, BRE Good Building Guide 38. How to calculate roof gutter size, with useful advice on design and installation of gutters.

Energy and carbon implications of rainwater harvesting and greywater recycling. Environment Agency Report: SC090018. Available from EA website.

Environment Agency Position Statement. The use of rainwater harvesting systems. GEHO0611BTYB-E-E. Available from EA website.

Rainwater harvesting: An on-farm guide, Environment Agency. Available from EA website.

Rainwater harvesting: Environmentally beneficial for the UK? Way, Martinson, Heslop & Cooke (2010). Water Science & Technology: Water Supply, 10(5), 775-781

Reclaimed Water Systems. Information about installing, modifying or maintaining reclaimed water systems. WRAS information and guidance note. 9-02-04.

Water Regulations Guide, WRAS, 2000 – guidance from WRAS (water regulations advisory scheme) on the Water Supply (Water Fittings) Regulations 1999.

Water Supply (Water Fittings) Regulations, 1999 – regulations that include means of preventing backflow from rainwater systems into drinking water systems.

Websites

Navitron forum (website: www.navitron.org.uk/forum). A good website for DIY minded people attempting to install RWH systems according to their own design.

The Meteorological Office (www.met-office.gov.uk) has rainfall data for the UK (for a fee).

Wunderground website: free rainfall data from many (unofficial) weather stations in the UK. www.wunderground.com

Chapter Seven

Grey water recycling systems

Why recycling your water isn't necessarily a good idea – what water might you recycle? – how might you recycle it? – combining sources of recycled water

The preceding chapters have suggested that we are going to have to go to quite considerable trouble to find our water supply, to clean it prior to use and to distribute it around the home. You might conclude from this is that we should use it as many times as possible before passing it back into the hydrological cycle.

You may have heard about water recycling systems and be aiming towards a completely 'closed loop' where you treat your dirty water from toilets, showers and so on, and make it clean enough to reuse. It is technically possible to do this, but you should certainly not consider this to be some ecological Holy Grail; rather the opposite, as you will see below. In the UK recycling domestic water for reuse within the house is rarely environmentally friendly.

You will remember from Chapter One that the first thing to do if you are aiming for a sustainable water system is to implement water efficiency measures. One result of this will be to reduce your dirty water production, hopefully to the point that the quantity of dirty water may be too small to justify a system for recycling it. Recycling water you have already used is not appropriate if you have a mains supply or access to a spring, well/borehole, surface water or rainwater supply, as these will probably be both more plentiful and of better quality (as discussed in Chapter Two).

In situations of extreme water scarcity, where your primary water source is insufficient to use for anything other than drinking, cooking and washing, then you may consider recycling for some of your non-potable water uses. Given that your main non-potable use is likely to be your toilet, you should consider minimising this use by installing a compost toilet as opposed to a grey water recycling system.

The type of sewage system you have also has a bearing on water recycling. If you are not connected to mains sewage and are about to build your own system, then you could certainly design it to include the potential for reuse of the final treated effluent in the garden. This type of system is discussed in *Choosing Ecological Sewage Treatment* (CAT Publications).

Unfortunately, the existence of 'off the shelf' grey water recycling systems as a product has become a clear cut example of where supposedly 'environmental' regulations (such as the Code for Sustainable Homes) diverge drastically from common sense. Simple calculations indicate that grey water recycling is unlikely to be either environmentally or economically sound. The available systems are also complex, require regular maintenance and suffer from water quality issues. Nevertheless, systems continue to be installed, so this chapter considers what would be the 'least bad' way of doing it. These systems are governed by a British Standard (BS8525), which includes guidance on design, installation, risk assessment and monitoring.

Which water sources are worth recycling?

With all water sources you have to consider the following factors:
- Amount available (and how regular this amount is).
- Quality (potential to contain contaminants).
- Type of use to which the water is going to be put (toilet flushing, garden irrigation or all domestic purposes).
- Ease of getting the water.

As you will see from table 7.1, the least dirty and most abundant of these 'waste' water streams in the home is the water arising from baths, showers and washbasins. Toilet water is just too 'dirty', water from the kitchen sink contains fats and food, and washing machine water contains high levels of detergents. The water from baths, showers and washbasins is often known as *grey water* and is potentially contaminated with:
- Hair and skin.
- Fats and grease from washing.
- Food waste.   } Nutrients
- Dirt and grit.
- Human faecal matter.
- Urine.
- Animal faecal matter (from pets or people handling them).   } Pathogens
- Detergents, soaps, surface cleaners.

Source of water	% total domestic water use (see figure 1.7 on page 19)	Regularity of production	Degree of contamination	Other points to consider
Toilet	35	Very	Very high	Compost toilet is likely to be more appropriate than recycling toilet flush water.
Washing machine	12	Variable depending on household	Wash: high Rinse: low	Difficult to separate wash water from rinse water. Washing nappies will significantly increase pathogen load.
Kitchen sink	15	Very	High	Fats and grease cause particular difficulties.
Dishwasher	4	Very	Low	
Baths, showers, wash basins	28	Very	Low	
Car washing, gardening	6	Irregular	Car washing: high	Collection is difficult as is on ground surface

Table 7.1. Sources of dirty water in the home and factors affecting their suitability for reuse. BS8525 recommends the following order of preference; showers and baths, wash and hand basins, washing machines.

Since grey water contains bacteria and a nutrient source, and is often discharged warm, you have an ideal situation for pathogens to multiply. Studies of installed systems in the UK have demonstrated coliform results in excess of 1000/100mls. Consequently if you are going to install a grey water system that recycles water for use in the house it is essential to have stringent safeguards to minimise the risks of cross-contamination.

Methods for recycling grey water

BS8525 categorises grey water systems according to the type of treatment they use:

1) Direct reuse – no treatment or storage, grey water reused in the garden (see Chapter Eight on garden watering).
2) Short retention systems – simple filtration, skimming of solids from surface or settlement. Deterioration of water quality minimised by avoiding long periods of storage.
3) Basic physical/chemical systems – similar to short retention systems, but with the addition of chemical disinfectant such as bromine or chlorine so that storage times and volumes can be increased.
4) Biological systems – use bacteria and aquatic plants to digest organic material in the grey water.
5) Biomechanical systems – similar to 'package treatment plants' for sewage treatment, often including membrane treatment.
6) Hybrid systems – combinations of the above.

Systems in categories 2 and 3 are typically available off-the-shelf, although the choice of models imported into the UK is still fairly limited. Biological systems and biomechanical systems (categories 4 and 5) are basically sewage treatment systems, and are discussed in *Choosing Ecological Sewage Treatment*.

Sizing the system: As indicated in table 7.1, the amount of grey water produced by a household (in other words the amount of clean water used in showers, baths and wash basins; about 28% of total domestic water use) is similar to the amount of water required for toilet flushing (approximately 35% of total water use). In other words, the supply to demand ratio is good, so grey water storage tanks can be fairly small (often around 100 litres for a domestic system) as they are replenished as regularly as they are emptied.

Two methods for sizing grey water recycling systems are given in BS8525; one for single household systems, and one for communal systems. The method for single households is based on an assumed 25 litre/person/day demand for grey water (for WC flushing), and a grey water production of 50 litres/person/day supply.

Final water quality: The water quality resulting from commercially available grey water recycling systems poses a low risk to health if the system is working correctly. However, more abstract indices of water quality such as odour (for example from the disinfectant), colour and cloudiness are often poor. Some systems include disinfectant with a deliberate colour, to provide a visual check for the user that disinfection has occurred, to mask the cloudiness and to distinguish the water from potable supplies. Notices alerting users to the presence of a grey water system and the impact this might have on water aesthetics should be considered in order to reduce the tendency of many people to flush toilets before use if the water in the bowl isn't clear.

Safety features: There is a real risk to health with grey water recycling systems if the treatment systems fail, or if the pipes are mistakenly cross-connected into other parts of the household plumbing system. Consequently they should have a variety of safety mechanisms integrated into them. These can include:

- Automatic drain down and discharge to sewer of the storage tank in the event of the system not being used in 24 hours. Whilst good for safety this reduces the volume of water available for reuse.
- Disinfection systems that fail in a deliberate way (for example they prevent the supply of grey water for reuse if the disinfectant runs out); again, good for safety but will reduce the volume available for reuse.
- Initial disinfection of water entering the tank.
- Separation of the inlets to the header tank to prevent backflow of grey water into the potable water supply.
- Artificial colouring to the water to discourage use for other purposes.
- Pipe labelling to prevent inadvertent cross-connection of grey water into the potable water supply in the house.
- An overflow to the sewer that is protected from backflow.

Individual grey water systems vary considerably in design and also in how many of these safety features are present.

Does domestic grey water recycling have a future in the UK?

Domestic grey water systems will not pay back economically (estimates suggest that the break even point would occur when mains water was £10/ m³, which is unlikely to be within the lifespan of anyone reading this book). Given the infrastructure requirements and the need for disinfectants or other treatment technology, grey water recycling systems cannot be considered environmentally friendly. They are also far from a 'fit and forget' system; annual servicing (with it's associated costs) is generally advised and BS8525 details the types of checks necessary.

The value of recycling water with high coliform counts is extremely dubious from a social perspective, and many people will find the aesthetic water quality in such systems unacceptable (for example cloudy, or liable to include foam or scum due to the presence of soaps and detergents).

In new build isolated properties where there is no mains water or drainage and other water sources are scarce, some grey water recycling may be appropriate, if implemented as part of a sewage treatment system involving simple filtration (for example by a reed bed or sand filter outside the house), rather than using one of the commercially available grey water recycling systems described above. In private sewage treatment systems where leachfield overloading is a problem, diverting grey water into a recycling system may improve sewage treatment. Nevertheless, a compost toilet would normally be preferable to recycling grey water for toilet flushing.

Despite the problems of grey water recycling systems, their use is likely to increase, particularly in environmental new build houses, owing to the fact they can be used to meet the water efficiency requirements of the Code for Sustainable Homes water calculator (see page 18).

Variations on grey water recycling systems: commercial buildings, community scale systems and combined rain/grey water systems

Some light industrial or agricultural enterprises exist where there is a high demand for non potable water, and also large volumes of grey water produced. Economies of scale dictate that grey water recycling may be justified in some of these instances (for example garages with automated car washes, fire stations, or vegetable washing plants).

Water treatment systems such as those described in Chapter Four are likely to be necessary. Standard commercial and public buildings are less likely to meet these criteria; daytime only use of a building usually means that grey water production is too low to justify a recycling system. There is a conflict with the water requirement for the Code for Sustainable Homes here; in low rainfall areas a rainwater harvesting system cannot meet a significant proportion of the toilet flushing requirements, and grey water production is also low.

Grey water systems serving groups of houses with centralised collection, storage and treatment have been installed at a number of sites in the UK. In these instances the filtration and disinfection procedures and management regimes are more elaborate than in single-household systems, as the risk

Combination before treatment	Combination after treatment
Only one collection tank and pump needed	One tank and pump required for each water type
High environmental and cost impacts of treatment since whole volume must be assumed to be from dirtiest source	Lower treatment impacts
Potentially leads to rain water overflow to foul sewer, highly unsustainable	Rain water tank can be allowed to overflow to surface water drainage since it contains no foul water contaminants
Mixing of source water does not allow categorising use according to source	Allows different qualities of water to be used for different purposes

Table 7.2. Combination rainwater and grey water systems.

to public health from poorly performing systems is greater. Whilst the economies of scale favour these systems over single-household systems it is still difficult to see how they can be considered sustainable at the present time.

Installing an integrated system that combines rainwater and grey water is also possible. This has the potential to increase the volume of reclaimed water available, with little additional infrastructure cost (at least in theory). Systems can combine the two water sources either before or after treatment, and as indicated in table 7.2, combination after treatment is likely to be more sustainable.

Combined rainwater and grey water systems require site specific design and are not yet widely available, nor have they proved themselves robust and cost-effective either domestically or on a community or commercial scale. An additional problem with these systems is what happens to any overflow from the tank. Since rainwater is not allowed to be diverted to mains sewers, special provision would need to made for conditions where the combined rainwater/grey water tank was overflowing. More detail on combined systems is given in BS8515.

Summary

Recycling your grey water so that you can use it to flush your toilet is not justifiable on environmental, economic or social grounds. Whilst at first sight it might appear to be efficient to implement a 'closed loop' water recycling system, we cannot consider water in isolation from other environmental problems. Grey water recycling systems typically require energy and disinfectants with higher environmental impacts than can be justified from the amount of water that is saved. Because of the approach to water efficiency taken under the Code for Sustainable Homes the use of grey water recycling systems is likely to increase. If you want to make use of grey water, then using it in a simple garden irrigation system, as described in Chapter Eight is far more sustainable than a commercially available system. In properties with private sewage treatment systems some separation of grey water may also be appropriate.

Further reading

BS 8525-1:2010 Greywater systems – code of practice. BS8525-2:2011 Greywater systems – requirements and test methods.

Choosing Ecological Sewage Treatment, Moodie, Grant and Weedon, CAT Publications, 2012.

Also see Further reading section from Chapter Six, since much of the literature includes both rainwater and greywater systems.

Chapter Eight

Water in the garden

How much is used – techniques for using less – simple reuse of rainwater – simple reuse of grey water – irrigation system design – coping with hosepipe bans

Water use outside the house accounts for just 6% of domestic water use in the UK, in other words nine litres per person per day, but is often the focus of water efficiency campaigns. There are several reasons for this. Firstly, water use outside peaks precisely when and where water supplies are at their most stressed. Summer demand for water is a third higher than in winter, and this is largely attributed to garden watering. It necessitates massive reservoirs to store winter water for use in the summer. Secondly, the use of hosepipes and sprinklers by large numbers of households simultaneously can result in decreases in supply pressure and subsequent problems in the distribution network; up to 70% of water demand on summer evenings can be due to use in the garden. Thirdly, water use in the garden is increasing, and this is expected to continue due to increasing affluence and interest in gardening, and the effects of climate change (which is expected to lead to wetter winters, but dryer summers in the UK). It is also regarded as a fairly non-essential use and therefore a relatively easy method of managing water demand, with assumed knock on effects on other water uses via awareness raising.

Water in plants has a number of roles. It is important for plant structure, takes part in many chemical and physical processes within the plant and acts as a solvent – allowing plants to take up nutrients. Water is continually flowing through plants; it enters through the roots, travels up through the plant and is then lost through the leaves via a process called *transpiration*. In addition to plants losing water by transpiration, water is also lost directly from the ground by evaporation. The rate at which this occurs will affect how much soil moisture is available to the plant.

It is possible to calculate the rate at which evaporation and transpiration occur, but since these rates vary according to species, size, humidity, soil

Water whenever soil seems dry (at a depth of 30cm)	Water every 10-14 days in dry periods	Never water
Perfect lawns	Smaller fruit trees and currant bushes	Everyday lawns
Leafy salads	Vegetables where the fruit is eaten (courgettes, squash)	Wildflower meadows
Peas and beans in flower	Bedding plants	Rough grass
Newly planted items		Large fruit trees
Containers and hanging baskets		Root vegetables
		Perrenial borders and ornamental grasses
		Drought-resistant succulent plants

Table 8.1. Watering frequency guide (courtesy of the Royal Horticultural Society).

moisture content, soil type, temperature and light levels it is difficult to water plants in a very efficient way using tables of theoretical transpiration and evaporation rates. Much better is to inspect the soil at a spades depth. If it feels damp, there is probably no need to water anything. A general guide to watering frequencies is given in table 8.1.

Avoiding the need for watering

The basic cornerstones of water efficient gardening are to choose the right plants, add plenty of organic material, and use mulches wherever possible.

Planting scheme

It sounds obvious, but you should try to grow plants that are suited to the local microclimate. If a plant in your garden seems particularly prone to withering away every summer, consider replacing it with something more suitable, or just see what appears. This isn't about limiting your choice of plants, it's more that the plants that are most drought tolerant will look best, as they'll inevitably be healthier than plants that you need to spend a lot of time looking after.

Try and rely on plants whose foliage and overall form look good, as this will always look better than wilting flowers. Drought tolerant plants often have pale, silvery leaves, and are sometimes hairy, leathery or waxy. Bear in mind that you will have different micro-climates within your garden; some areas will be more shaded from sun and wind, and therefore suffer less from evaporation.

If you have favourite plants that don't survive well through the summer, make sure they're planted close to the water butt so that they are easier to water, or put them somewhere where they will be easy to irrigate with grey water. Plant new plants in autumn or spring so they have a good chance of growing roots before a dry summer. Several books consider plants suited to dryer parts of the UK (see Further reading at the end of this chapter).

Organic material

Whatever your soil type, it will be vastly improved by the addition of compost. Organic material will improve the texture of soil; if you have clay soil it will help open up the structure and improve drainage, and, at the other extreme, if you have a sandy soil it will act as a sponge to retain moisture. In both cases the organic matter will make it easier for your plants to withstand periods of drought. According to the Royal Horticultural Society (RHS), adding compost to soil is equivalent to an extra 5cm of rainfall. You should dig in to the soil as much compost or well rotted manure as you can, but if that sounds like too much effort you can simply spread it on the soil surface, or put it in trenches to target it to specific plants.

Mulches

Mulches are essential to water efficient gardening; they slow evaporation so that plants don't need watering as frequently, and they also suppress weeds (which use water that could otherwise be used by your precious plants). There are two basic types; organic mulches (figure 8.1a) and sheet mulches (figure 8.1b).

Any organic material that will compost is suitable, such as wood chips, bark, coir and semi-composted kitchen waste. The loose structure of organic mulches means that any weeds that do appear are easier to pull out, and the mulch provides a good habitat for invertebrates (including, unfortunately, slugs!). As the organic mulch breaks down it will increase nutrient levels in the soil, loosen the structure, and improve water-holding ability.

Sheet mulches can increase the surface water run off if they are completely impervious, but woven and porous sheet mulches are also available. When used in planted areas where slits are cut in the sheet material (for the plants to emerge through) the sheet mulch helps direct any rainfall directly to

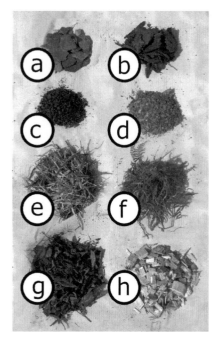

Figure 8.1a (left). Organic mulches. (a) pebbles (b) bark (c) cocoa (d) sawdust (e) straw (f) grass (g) leafmould (h) wood chip.
Figure 8.1b (above). Sheet mulches. (a) sack (b) geotextile (c) carpet (d) cardboard.

the plants. In terms of sustainability, reused or recycled materials are preferable for sheet mulches, such as old carpets (not the foam backed ones), or cardboard (since this will eventually rot down to compost, it is a cross between an organic and a sheet mulch). Since these can often look messy you may want to put organic mulch on top of the sheet mulch for aesthetic reasons.

Lawns, trees and ponds

Do you really require your lawn to be green and lush, or can you put up with it going a bit brown for a few weeks in the summer? A happy medium may be to reduce its area, or replace it with some other surface (preferably an eco-friendly one like vegetable or ornamental beds; not concrete or tarmac!). Increasing the organic content of the soil will help it stay green; organic matter retains water so top-dress the lawn every year with compost. Other tips:

• Long grass will stay greener longer than short grass, as it provides shading for the soil and lower leaves, and leaving cut grass on the lawn surface will allow it to act as mulch and help retain moisture.

- Compacted lawns will tend to let surface water run off more quickly so try and keep the top layer fairly open by spiking and then brushing compost into it.
- If you're re-sowing your lawn, try and get seed that is drought-tolerant. Specialist mixes are available that contain high proportions of grasses such as creeping red fescue. New lawns should be planted in the autumn, and you should take care to make sure the soil contains lots of organic matter to help retain water; try and spread at least 5cm of compost over the entire area before you seed the lawn.

To help saplings through their first few years, you can sink a pipe a couple of feet into the ground next to them, so that you can easily apply a large volume of water, without allowing much run off or evaporation. Once a tree is established it shouldn't need any extra watering (and you don't want to encourage surface roots by watering them).

If you have ponds or water features in your garden, consider how you can make them more water efficient. You can cut down the evaporation rate by allowing surface plants to cover a large proportion of the water surface. This will also help prevent it heating up (and since less oxygen is dissolved in warm water, the fish will prefer it cooler as well). This may in turn decrease the need for aeration and mixing (both of which tend to increase the evaporation rate). Pond life prefers rainwater to mains, so refilling your pond from your rainwater butts has dual benefits.

Vegetable gardens

If you grow your own vegetables you will need to keep them well watered, but you should try to group them according to their water needs. Root crops are the most drought tolerant. Courgettes and squash may only need extra watering once the fruit starts to swell. Cauliflower, broccoli, salad onions, peas, potatoes, runner beans, tomatoes and leafy vegetables will need the most water, so should be put together somewhere where it's easy to water. Whilst this makes sense from the water efficiency perspective, it is not necessarily possible if you are carrying out crop rotation. Also bear in mind that you can delay plant maturity and usefully spread the harvesting period by not watering some areas of the veg patch.

Using water efficiently in the garden

The above measures will help make your garden as water efficient as possible, but you may still need to water your garden sometimes. There are methods for improving the efficiency of watering, and you may also be interested in water other than that from your mains (or private) supply. If you are gardening commercially or on a large scale, you should refer to Chapter Two, which describes how to go about accessing large ground and surface water sources. This chapter is aimed more at basic techniques for the irrigation of domestic gardens.

A garden sprinkler can use between 600 and 1400 litres an hour (compare this to the water uses illustrated in Chapter One). Even when considered in terms of the water per square metre, this is far more than any plant needs, and most is lost to evaporation. Sprinklers also encourage plants to form roots near the surface, which will die off more quickly during dry periods. However, sprinklers remain the simplest and most commonly used technique for watering lawns. If you can't live without yours, get an electronic timer so you can water for a short period during the night, when evaporation is lower and demand on the mains network is reduced. The timer will allow you to programme when and for how long you water and the sprinkler will be turned off automatically by a solenoid valve.

If you can't carry watering cans and need a hose to water in different areas of the garden fit a trigger nozzle so you're not wasting water whilst moving from one area to another. Avoid watering little and often (as this will encourage surface roots). Watering every seven to ten days is preferable (and less work). This is actually quite difficult to achieve. Because water won't soak into the ground very rapidly it can be quite time consuming to give a plant enough water without most of the water running off the surface. A flow restrictor or nozzle that gives a variety of options for flow will help; flow rates from hoses are usually much too high for easy watering, particularly of hanging baskets and tubs. Another tactic is to sink an empty plant pot beside each plant to water in to. This allows the water to get to where it's needed without much evaporation or run off, and allows you to see how much water you're giving the plant (figure 8.2). You can also create small basins in the earth around plants, or raise the earth a little on the downhill side of plants in a sloping garden to help to slow run off of surface water when it rains (figure 8.3).

Considering other water sources

The sources available to you for garden watering are likely to be mains water, rainwater and grey water. Whilst it is wasteful of energy and resources to use drinking quality water in the garden, some care is needed if you are going to

Figure 8.2 (left). Sinking an empty pot in the earth next to a plant allows you to tell how much water you've given it and reduces the amount that runs off the ground surface. Figure 8.3 (above). A small depression around each plant minimises the surface water run off when it rains.

use lower quality water. This is particularly the case if you are watering food crops that you will then eat without first cooking them. Spray or sprinkler irrigation also poses a risk, since you may breathe in the water as airborne droplets, or swallow it. The following basic rules are offered as guidance for use of water in the domestic garden:

- Neither rainwater nor grey water should be used in sprinkler or spray systems without prior disinfection.
- Grey water should be applied below the soil surface wherever possible, and should never be stored for periods longer than 24 hours unless it has been treated.
- You should avoid using grey water for irrigating crops that may be eaten raw, particularly if your grey water includes washing machine water from nappy washing.

Simple rainwater collection for garden watering

There are lots of environmental reasons why it makes sense to collect rainwater for use in the garden:

- Water will be returned to the water cycle in the catchment area in which it fell, rather than being collected from a network of properties and discharged as stormwater elsewhere.
- Dispersing rainwater to the ground close to where it falls means it won't get mixed with the dirtier water that runs off roads and paved areas.
- Decreased pressure on drainage infrastructure. This is particularly important in areas with a high proportion of combined sewers, which

can discharge raw sewage to rivers in the event of heavy rainfall.
- Plants don't need drinking quality water, they will actually do better with rainwater than mains water.
- Reduces demand for drinking quality water in water scarce areas.

Unfortunately, the time when your garden needs the most water is usually the time during which there is least rain. So we immediately have a mismatch between supply and demand, which as you will remember from Chapter Six, means that large storage volumes would be necessary to meet your entire demand. In reality, in most of the Southern England, you will not be able to store enough winter rainfall to meet your summer gardening needs unless you store thousands of litres. Nevertheless, in most instances doing something is better than doing nothing. There are several levels of complexity possible, so you should decide fairly early on in the process how much effort you wish to go to.

1) Put water butts at the bottom of all rainwater downpipes. For many people, the limiting factor to how much rainwater they can store is determined by the practicalities of siting water butts. Slimline rainwater butts are available if you are short of space, but they obviously hold less water. For high capacity systems consider intermediate bulk containers (IBCs); they are cheap, widely available, can be stacked easily to increase capacity and provide enough head to allow direct garden watering (figure 8.5).

2) Make some simple modifications to your water butts (figure 8.4). Put a tap in the bottom of the water butt, and put the butt on a strong platform so you can get a watering can under the tap. An overflow pipe diverted to the existing rainwater drain will prevent the water butt overflowing from the top and causing dampness in any adjacent wall. It will also make it easy to assess whether or not it is worth increasing your storage capacity – by seeing how often it overflows. It's always worth putting a lid on your water butt, even if it's just a piece of wood. A lid improves safety, decreases evaporation, prevents contamination and helps stop potential pathogens breeding in the water by minimising exposure to light. It also helps prevent mosquito larvae breeding. These issues all become more important as storage volumes (and therefore storage durations) increase.

3) If your garden topography allows it, run a hosepipe from the tap at the bottom of a water butt to allow you to water the garden directly. The water pressure will be better using a tall water butt.

4) Rainwater filters and kits that divert water from your downpipe both remove solids like leaves and twigs from the rainwater. Crucially, they

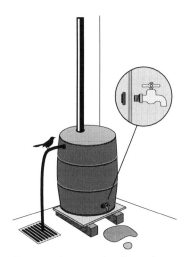

Figure 8.4. Rainwater butt for garden watering.

give you more flexibility in water butt placement. Removing these solids is also a good idea if you are planning an irrigation system, so that the holes in your irrigation pipe don't get blocked. If you're just filling watering cans, filters and diverter kits may not be worth the expense.

5) **If you can calculate your garden water demand** then you can do the maths to work out how much water you should store. Refer to the calculations in Chapter Six, which will help you optimise your collection system.

6) **In a domestic new build situation, or where you are having landscaping work done,** you might consider large underground water storage tanks. You could use this stored water for toilet flushing in the winter and garden watering in the summer.

Reusing grey water in the garden

The mismatch between supply of rainwater and the demands of garden watering means that in some circumstances, grey water is potentially a useful addition. As discussed in Chapter Seven, grey water is the waste water arising from showers, baths, basins, washing machines, sinks and so on (in other words all household waste water except that from the toilet). If you are following the principles of reduce, reuse, recycle, you will already have implemented water efficiency measures and your grey water production may be around 20-30 litres per person per day (as opposed to the UK average of around 100 litres). You should think carefully about how much effort it is worth going to in order to reuse this fairly small volume of water. In addition to its low volume, grey water is often dirty and therefore difficult to deal with. You should certainly put effort into a rainwater system before concerning yourself with grey water. However, grey water is produced all year, so, unlike rainwater, it isn't going to run out exactly when the garden needs it most.

Box 7.1. Case study – watering a vegetable patch and greenhouse

Rainwater is collected from the roof of a barn (90m²); local rainfall is approximately 740mm/year. The guttering leaked and needed replacing, so was replaced with steel guttering. The opportunity was taken to re-route the guttering from both sides of the barn to one point. A downpipe filter is used to prevent leaves and other debris entering the tanks. The IBCs (intermediate bulk containers) cost £65 for the pair; excellent value for 2000 litres of water storage. The height of the tanks above the ground and the use of 25mm pipe from the outlet to the garden tap means that there is sufficient head for a reasonable flow rate through a garden hose. The system has supplied sufficient water for garden use in the summer, but could easily be expanded with another IBC if required. The system is drained over winter in case of freezing.

Figure 8.5. Intermediate scale rainwater collection for garden watering.

How 'dirty' is grey water?

Some of the things that make grey water 'dirty' – such as soaps, shampoos, washing powder and cleaning agents – may not be good for your soil. The components which are the main cause of concern are sodium, pH, phosphorous, and solids and fats.

Sodium: Sodium is an essential nutrient in the soil, but if levels are too high, the effect on your plants is similar to that of drought. The most significant source of sodium in your waste water stream (around 40%) is washing powders. The sodium is present as a filler or bulking agent, does not contribute to wash quality and yet can be the equivalent of pouring 90g of table salt down your drains every wash! Low sodium laundry detergents are available; these are typically concentrated liquid detergents or tablets as opposed to powdered products. Similarly, you should not use softened water in your garden, or the backwash water from a water softener, as this can also contain high levels of sodium.

pH: The pH of household cleaning chemicals, soaps and detergents tends to be very alkali, since grease and soiling is easier to remove at high pH. Plants that like acid soils may not thrive if irrigated with alkaline grey water. High pH in wash water can also reduce the life of clothes and cause skin irritation, as well as damage soil structure.

Phosphorous: Phosphorous is an essential plant nutrient, but too much of it is a bad thing, particularly in the aquatic environment where it results in eutrophication, a process in which algal blooms arise in response to the increase in nutrients and oxygen levels fall. So, whilst it won't cause your garden any problems, phosphorous has wider negative environmental impacts. The phosphorous load in sewage comes largely from detergents, and since compounds called zeolites can replace phosphorous in laundry detergents, it is unnecessary; in many countries its use is banned.

Solids and fats: Grey water contains varying amounts of solids, such as skin and hair, and kitchen sink grey water contains a lot of food solids and fats. These aren't bad for the garden, but they will tend to block pipework.

Simple grey water reuse

In most circumstances, grey water irrigation is best if it's kept simple. You may not want to reuse grey water from certain appliances, and it may be something you set up just for the summer. The main exception is if you don't have a mains sewerage connection, when a more elaborate system may be appropriate or even necessary. Grey water reuse can reduce the load on your septic tank, or with a more detailed design can be a treatment system in its own right. Waste water treatment often includes a stage of filtering through gravel or sand in a tank; plants can be grown in this filter medium, thereby combining gardening and sewage treatment. If you have a compost toilet, grey water reuse will mean you may have no need for a formal sewage treatment system at all. Interested readers are referred to *Choosing Ecological Sewage Treatment.*

At its simplest, grey water reuse consists of emptying the washing up bowl into the garden. There is much to recommend this minimalist approach; it has no construction costs, allows irrigation exactly where it is needed, and

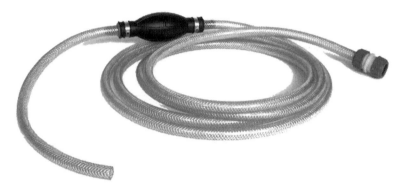

Figure 8.6. 'DroughtBuster' siphon.

requires no maintenance. If you want to drain your bathwater out into the garden on an occasional basis, it's not really worth modifying your plumbing if you can simply drape a hose out of the bathroom window.

You can set up a siphon without getting a mouthful of bathwater by using a 'droughtbuster' (figure 8.6) or a jiggle siphon (most often used for siphoning fuel, so look them up in automotive supply catalogues). Compare the average bath full (80 litres) to standard sized water butts (200 litres) and you can see how this simple technique can help your garden through the summer when your rainwater butts are empty. Shower trays are fiddly to re-plumb and don't hold much water, but if your shower is above the bath, leaving the plug in whilst showering allows this water to be reused.

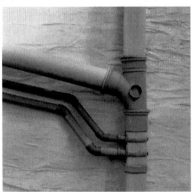

Figure 8.7. In properties with external sewage pipes (usually 110mm diameter), the pipes carrying grey water from the baths and washbasins (usually 32 or 40mm diameter) may be fairly easy to disconnect and divert into a grey water irrigation system.

The next level of complexity involves some plumbing: separating your grey water waste pipes from your main soil stack. If you have a toilet upstairs, there will be a 110mm wide pipe carrying the toilet water to the main sewer beneath ground floor level. Smaller (32mm or 40mm) pipes carrying grey water from baths, showers, basins etc will be connected into this pipe at various points.

Figure 8.8. 'Water Two' device incorporating a valve (operated manually by pulling the strings) to divert grey water.

These pipes can be located either inside the house in a duct, or outside the house attached to a wall. If the soil stack is on an external wall, then it will be easy to tell where the grey water joins, and so reconnect it into a grey water system (figure 8.7). You can either make this reconnection permanent, or use a 'Water Two' device (figure 8.8). This incorporates a valve operated by pull cords to divert the grey water. If you have an internal soil stack, diverting the grey water might require drilling through external walls, which could be quite a lot of bother in proportion to the small volumes of grey water produced by a water efficient household.

A simple system for grey water from washbasins

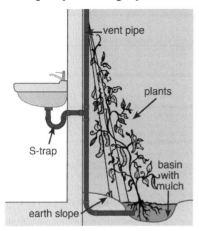

Figure 8.9. A simple grey water reuse system for washbasins.

Wash basin waste water is produced in small quantities and can be reused using the system illustrated in figure 8.9. The outlet pipe can travel any distance from the house, provided it has at least a 2% fall. Depending on the layout of your garden, it may be possible to have several of these systems, one for each wash basin. This system isn't so good for kitchen sink water, which contains a lot of solids and grease and which will tend to block any pipe-work.

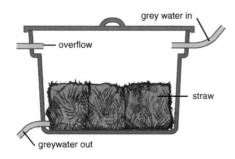

Figure 8.10. Grey water surge tank to filter grey water for garden use.

Whole house grey water systems

If you wish to collect grey water from the larger volume sources (such as showers, baths, washing machines), you will need to incorporate a simple filter and surge tank. The system shown in figure 8.10 will: remove solids (and so minimise the risk of blocking any irrigation systems); allow the water to cool; temporarily store some of the water so that household appliances drain quickly. A tank of around 40 litres will be sufficient, unless you have a particularly high grey water output.

Having filtered the grey water and evened out the flow a bit, you need to consider how and where to distribute it in the garden. As stated earlier, grey water should be used within 24 hours, because it will quickly stagnate and become smelly. A system that automatically discharges into the garden is preferable to one in which handling is required. It is also better if applied below ground level in a sub-surface system; *on no account should untreated grey water be used in spray form*. Design of simple irrigation systems, including ones suitable for grey water is described below.

Domestic irrigation systems

Irrigation systems with more permanent networks of pipes are classified according to how the water is applied, and vary in their suitability for different water sources, as indicated in table 8.2.

Overhead mist systems: These consist of pipes held above ground level, dispersing water in mists or droplets. They are often mounted in polytunnel or greenhouse roofs.

Sprinkler systems: They can be as simple as the common lawn sprinkler you attach to a hose or more complex, for example with a network of pipes and pop up sprinklers buried underneath the lawn. You can even get 'walking' sprinklers that move themselves across the garden.

Overhead systems and sprinklers are incredibly inefficient, as they allow massive amounts of evaporation. This is very temperature dependent, but can exceed 80% of the water applied under UK climatic conditions.

System	Water efficient?	Water source irrigation technique is appropriate for		
		Mains	Rain	Grey
Overhead mist	No	Yes	No	No
Sprinkler	No	Yes	No	No
Surface drip – commercial	Yes	Yes	No	No
Surface drip – DIY	Yes	Yes	Yes	No
Sub-surface	Yes	Yes	Yes	Yes

Table 8.2. The type of irrigation system you use should depend upon what type of water you want to use in the system.

Additionally, neither overhead nor sprinkler systems are suited to rain or grey water that hasn't been disinfected, since they result in airborne water droplets (and hence the risk of breathing in contaminated water). Because of this they will not be considered further.

On-ground (or surface) drip system: These are basically pipes with holes in that allow water to drip onto the ground surface. Commercially available irrigation pipe for drip irrigation usually has very small holes. There are two types; in *soaker hose* they are so small that the pipe appears to 'sweat', whereas *trickle pipe* is a thin collapsible polythene tube with slits in that larger droplets come out of (figure 8.11). Little water is lost by evaporation,

Figure 8.11. Trickle tape. This thin collapsible tape is an easy way of irrigating the garden. It is ideal for mains water irrigation systems, but because the holes in it are small it is not well suited to use with rainwater or grey water systems (courtesy of Access Irrigation).

and losses can be reduced further by covering the pipes with mulch. Neither is suited to rainwater or grey water, since the holes will block easily. On-ground systems are very adaptable; the pipe can simply be moved from one bed to another.

Sub-surface irrigation systems: These consist of pipes with regularly spaced holes, layed in trenches that are then filled with mulch. They are a fixed installation (more so than 'on ground' drip systems which can be moved fairly easily), and are more work to install. However, they are more water efficient; no water is lost to evaporation, and water is applied directly to the roots of the plant rather than the ground surface. You can buy perforated pipe, or make your own by piercing holes in a piece of hose pipe.

As indicated in table 8.2, on-ground drip and sub-surface irrigation systems can be used with mains water. However, the latter are most water efficient and can be used with any water source, so are the best option for most gardens. The design process for a simple domestic irrigation system is described below.

1) Draw a plan of the garden and decide where you wish to water.

2) Rainwater only. Decide where you're going to store the water, and how you are going to get it from the tank to the area to be irrigated. The higher up you can store the water relative to where you want to irrigate, the better the system will perform. The relationship between water pressure and flow is complicated, but as a rough guide you will need to have at least a metre of height difference for every 5m of 16mm diameter dripper pipe.

3) Will it be an on-ground system or a sub-surface system? On-ground systems are simpler to install and make it easier to move the pipe around, but sub-surface systems are more water efficient and allow you to incorporate grey water as well as rainwater.

4) Work out the most efficient way of laying out the pipes. The water pressure will drop along the length of the pipe and so less water will be delivered through the holes further away from the supply end. You should therefore try and have a main pipe with branches, rather than a single length of pipe (figure 8.12). This design will also allow you to have taps on the branches, so you can concentrate the watering where it is most needed. Limiting the distribution to individual branches at any one time will also help if you have low water pressure. The end of each pipe should have a stopper in it to improve the distribution of water. You can remove this stopper periodically to flush water through the pipe-work to clear any blockages. Slopes will also affect the distribution (commercially available pipes have pressure compensation nozzles to reduce this effect, but the nozzles are liable to block if used with rain or grey water).

5) Work out how many connecting pieces you need and of what type (T pieces, elbows, blank ends and so on), and how much pipe you need. You may also need some specific plumbing parts like check valves or 'Type DB valves' to minimise the risk of any rain or grey water entering the mains water system (see page 187 under section 'Regulations that apply to water in the garden'). You can make your own grey water irrigation pipe out of

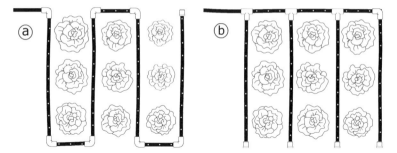

Figure 8.12. Pipe layout. Pressure will drop as water flows along an irrigation pipe, so it makes sense to have a main pipe with several branches coming off it (as in b), rather than a single continuous pipe (a).

Figure 8.13. On-ground (left) and sub-surface (right) irrigation systems.

hose pipe by piercing holes in it, or out of plastic waste pipe (32 or 40mm). Hose-pipe has the advantage of flexibility, so you won't need so many joints, but is more difficult to lay flat. There are no hard and fast rules on how big the holes should be and how far apart; it is very dependent on the length

Figure 8.14. Protruding distribution pipe for a grey water system.

and diameter of the pipe and the incoming pressure. At CAT we have used 32mm waste pipe with 8mm holes every 30cm; half inch hose-pipe with 3mm holes every 20cm also works well.

6) Install the pipe work. Sub-surface systems are laid in 10-20cm deep trenches alongside each row of vegetables in the garden (figure 8.13), and the trench backfilled with woodchip to act as a mulch. Using mulch filled trenches rather than laying the pipes directly into the earth reduces the risk of root penetration and soil blocking. It also allows pipes to be easily removed and cleaned should they become blocked. Over time the woodchip composts and thus needs to be replaced every couple of years. If you are planning a number of lines of pipe you can make the system more adaptable by allowing the end of the pipe to emerge above ground level, and simply put the end of the hose from the surge tank into whichever pipe run you wish to use at the time (figure 8.14).

Other components
Automatic controllers

It's all too easy to forget to turn off a garden tap, and lose all the precious rainwater you have been collecting for months, or waste hundreds of litres of mains water. A timer that will turn off the tap after a fixed duration is therefore a good investment. The more complex timers will turn water on

as well as off, and allow you to set a frequency and duration of watering (figure 8.15), but you should bear in mind that any device that turns the water on is likely to waste more water than it saves, unless you also remember to deactivate it during the weeks that the garden doesn't need watering.

Devices that turn off automatically after a set *volume* rather than a set time are also available, but owing to the way in which the switch works, they have a minimum operating pressure of around 0.5 bar, so are generally only suited to mains water irrigation systems. Under the Water Supply (Water Fittings) Regulations, prior to installing any automatic water system fed from mains water the local water company must be notified (on official notification forms).

Figure 8.15. Water timer. This device can be fixed to an outside tap to turn it on and off at times set by the user. This can reduce water wastage if you might otherwise forget to turn the tap off.

Pumping water for garden use

Electrically powered pumps suitable for moving hundreds of litres of water per day are dealt with in Chapter Five. If you want to pump small volumes of rain or grey water for the garden you can often do it by hand. For example, you may want to pump rainwater from a water butt at the front of the house to a more convenient one at the back. Hand operated bilge pumps (figure 5.12 on page 123) connected to garden hoses are well suited to this purpose, and are available from around £40. Most will pump to around 3 metres height and you can pump 40 litres a minute without too much physical effort. If you just need to set up a siphon, the droughtbuster (figure 8.6 on page 180) is your best bet.

Do not connect the outlet hose of your washing machine to a grey water system and use the power of the washing machine pump to move the water up a gradient into an irrigation system. These pumps are not designed to take this type of load and so if you want to use the grey water arising from your washing machine you should discharge it via a conventional standpipe and trap. This in turn may be connected to your grey water system if the gradient is suitable.

Regulations that apply to water in the garden

It is important to minimise the risks of contaminated water from the garden being sucked back into the mains water supply. This could happen if you are filling a water butt with a garden hose and the pressure in the mains water supply drops; water from the water butt could flow back into the supply pipes. Measures required under the Water Supply (Water Fittings) Regulations are designed to prevent this type of event.

Outside taps require a 'double check valve' either in the supply pipe (in new installations), or within the tap itself (where an outside tap is being replaced). Suitable taps will say 'HUK1, HA, EC or ED' on the box. These simply prevent water flowing the wrong way. Similarly, you shouldn't refill water butts from a hose attached to indoor taps since they won't have this backflow protection.

Irrigation systems connected to a tap require the tap to have a double check valve as described above, but also a 'Type DB device', which is a more failsafe backflow prevention measure. NB: More stringent precautions are necessary in commercial installations; you should consult the WRAS Water Regulations Guide, or an irrigation engineer if you are unsure what is required.

What to do during a hosepipe ban or drought order

Hosepipe bans and drought orders are a dreaded feature of summers in England. A hosepipe ban makes it illegal to use a hosepipe or sprinkler attached to a mains water supply tap for watering private gardens or to use a hosepipe for washing cars. They do not restrict your use of rainwater stored in water butts, or grey water, or prevent you from using a hosepipe that isn't attached to a mains water supply tap. Neither do they prevent you from watering with a watering can, or using buckets. You can use hosepipes to move stored water around (although you might want to point out to any nosy neighbours exactly what you're doing so they don't try and get you into trouble) and you can continue to use hosepipes/sprinklers from private water supplies such as boreholes, although it would be pretty irresponsible and un-environmental for you to do so. Note that if you have a commercial rainwater harvesting system that incorporates a mains water backup to the storage tank, you will not be able to use this tank of water; it will be mains water rather than rain.

Drought orders are more extreme than hosepipe bans, and the limitations on water use will vary, but will be well publicised by the water company concerned. In the most part, they don't affect householders but are aimed more at larger scale users. However, they will often include limitations on filling of private swimming pools and ornamental ponds.

If you've planned ahead for an impending hosepipe ban, you will doubtless have some rainwater saved in butts. However, most people won't be able to store as much as their garden needs, so you may well start using your grey water as discussed earlier in this chapter. Calculating how much water your garden actually needs during drought conditions will allow you to consider how to design storage and irrigation systems for future years.

Summary

There are a number of ways of making your garden water efficient, and additional water should be collected from roofs in preference to the use of mains water. DIY irrigation systems can be used to optimise the use of this water, and can be adapted to include use of grey water from baths, showers and washbasins. If you only really need to water your garden for a couple of months in the summer during a hosepipe ban, simple grey water recycling techniques should be sufficient.

Further reading

Garden watering systems, Susan Lang, Sunset Books, 1999. Irrigation system design and installation.

The Dry Garden, Beth Chatto, Orion Books. Advice on choice of species and garden design in dry areas.

Watering systems for lawn and garden, a do-it-yourself guide, R Woodson, Storey Books, 1996. Irrigation system design and installation.

Websites

www.access-irrigation.co.uk Fact sheets available on various aspects of garden watering. Suppliers of equipment suited to mains and rainwater systems.

www.hosepipeban.org.uk Hosepipe bans are well publicised, but this website collates information from all the UK water companies, and also includes water saving tips.

www.myclimatechangegarden.com Blog including lots of ideas for drought tolerant plants suited to the UK and where to buy them, useful gardening tips.

www.oasisdesign.net/greywater A wealth of information on grey water use in the garden (an American site).

www.rhs.org.uk/advice The Royal Horticultural Society has lots of useful information on water efficient gardening and drought tolerant plants. The RHS garden at Hyde Hall has an area dedicated to drought tolerant plants.

Conclusions

The largest single measure that can be undertaken to preserve the life and health of a population is to install safe water supply and sewage treatment systems. In the UK we have a safe, robust system that provides millions of litres of clean water a day to our houses, collects water we've made dirty, cleans it and then releases it back into the environment. Any system of this scale clearly has some environmental impact, and as responsible citizens we should do all we can to modify our behaviour to reduce this impact.

However, our impact due to water and sewerage is one of the lesser of many evils, comprising an average of just 2% of our ecological footprint. As discussed in Chapter One, our most significant water related impact is our use of hot water, as opposed to any impact that our local water company has. Consequently for most people whose primary motivation is the environment, it is not appropriate to prioritise water use above energy, transport, purchasing decisions or food miles. It would make far more sense to consider your *water footprint*; the water that is required for the goods and services you consume.

Nevertheless, there is a lot we can do. Since using less of a resource is easier than finding more, water efficiency should be the first resort and there are many behavioural changes that will help you reduce your water consumption. If you are upgrading water using appliances, you should bear in mind their water efficiency, and the advent of water efficiency labels on products makes this easier to determine. Collecting rainwater in butts for use in the garden is also a good strategy as it goes someway to alleviating flooding problems with surface water drainage, and also problems with droughts and hosepipe bans.

In new build properties, or those with larger roof areas, it may be worth looking at more elaborate rainwater harvesting systems, possibly including reuse for toilet flushing in addition to the garden. Hosepipe bans are a regular feature of the English summer, and planning ahead will enable you to go along way to preventing your plants from suffering too much. These measures may well include some recycling of 'grey' water from showers and baths.

The merits of installing a private or supplementary water supply will vary widely between households and we would urge you to be mindful of the bigger picture before deliberately disconnecting from, or supplementing, mains water.

However, in some locations, mains water isn't an option. Finding your alternative water source will be your first challenge. In some instances there will be a variety of sources available and the challenge will be the decision making process on which to use. In other instances, the chance of even finding a water source may seem remote. An experienced bore hole drilling company may be able to help you at this point.

Once you have found your water, you need to make it safe to drink, and avoid doing anything to contaminate the water source itself. The way in which you clean water depends on what is in it that you are trying to remove; it is best to start from some water test results rather than a preconceived idea about what type of treatment system you would like. In most small private systems, water testing is infrequent or completely absent, and the practical use of testing is fairly limited. It is far better to have a risk based approach to safeguarding supplies, as discussed in Chapter Three.

There are many reasons why you may wish to install a private water supply system, other than considerations of environmental impact. However it is not a decision to be taken lightly given the risk to your health from a poorly functioning system. No private water supply system will be 'fit and forget'; if you are not the type of person who is likely to get round to checking the system and doing maintenance when required, make sure you have an arrangement in place to make sure that it does get done. Hopefully by reading this book you will be in a better position to make an informed decision on suitable water sources, treatment methods and upkeep whilst maintaining an environmental perspective.

Appendix 1 –
Private Water Supply Regulations

As discussed on page 62, the Private Water Supply Regulations will apply to your water supply in certain circumstances. Even if they don't, the water quality parameters within them represent limits widely accepted by scientists, and other elements within the Regulations, such as risk assessment, are also helpful. They are implemented by the Local Authority Environmental Health team (or the Department of the Environment in Northern Ireland) who should be consulted if you are unsure about your specific case. The Regulations include provision for the local authority to recover their costs for carrying out monitoring and risk assessment under the Regulations, but the maximum amounts chargeable are fixed. There is also an official grant scheme in Scotland to help householders bring supplies up to standard. The local authority has powers of entry under the Water Industry Act, so refusing access to prevent risk assessment or sampling is not an option.

For any supply, the Regulations stipulate how the 'relevant person' to whom the Regulations apply is determined. In some circumstances this will be a number of people, and will generally include the person that provides the supply, anyone that has any control over it, the person owning the land where the supply originates from, and anyone owning or occupying premises which are served by it. The 'relevant persons' will be established by the local authority in consultation with people falling into the above groups.

As shown in table 3.2 on page 64, private water supplies are subdivided into 'large supplies' (called Type A supplies in Scotland) and 'other supplies' (Type B supplies in Scotland). Large supplies are further subdivided into a number of sub-categories based on the volumes of water involved, and these sub-categories are used to determine monitoring frequency. The aim of the classification scheme is to separate the widely differing types of supply into similar risk groups that can be considered together.

Note that any supply serving commercial or public activity is automatically assigned to the more stringent 'large supplies' category. This definition will therefore include bed and breakfast accommodation, remote camp sites and caravan parks, even very small ones. This reflects the fact that transient populations may have a higher susceptibility to poor water quality.

Risk assessment

The main difference between the PWS Regulations and the 1991 version they replaced is the requirement for risk assessment. The rationale for this is discussed on page 62. The local authority will do a risk assessment at least once every five years (they may increase this frequency if they feel it is warranted). The recommended form of the risk assessment is detailed on the Private Water Supplies website (www.privatewatersupplies.gov.uk).

Note that it is an entirely health based risk assessment; it is not intended to assess the quantity of water and whether the supply is at risk of drying up during the summer. The assessment is largely based on source control measures (table 3.3 on pages 65-68 is based on it), and assigns numerical values to risks of various types in order to weight them. The assessment will take into account both the severity of the hazard and the likelihood of it actually occurring in order to come up with an overall risk score. The way the assessment is constructed means that lots of supplies will be deemed high risk; this is partly because if the answer to any question is 'don't know', then a high numerical value is assigned.

Monitoring requirements

In addition to risk assessment, supplies will be sampled for a range of parameters. The parameters monitored differ between large supplies (where lots of parameters are monitored) and other supplies (where the presumption is that far fewer parameters are routinely measured, but additional parameters can be included if local circumstances suggest that it is necessary).

Large supplies (Type A in Scotland): Two types of monitoring are undertaken for these supplies; 'check' monitoring and 'audit' monitoring. Check monitoring is for a subset of 17 parameters that are either particularly common or dangerous, whereas audit monitoring is for a much wider set of (at least 30) parameters that reflect those monitored for in mains water supplies. Audit monitoring includes parameters that may not cause immediate harm but would pose a risk if ingested over a long period of time. The LA may choose not to include monitoring for many of these parameters if they do not regard them as a significant risk. The frequency of both types of monitoring is determined by the volume of water supplied. Consult the Regulations or your local Environmental Health team for details.

Parameter	Value	Comment
Conductivity	2500 μS/cm	A general measure indicating the potential for presence of certain contaminants. See page 90.
Enterococci	0/100 ml	A micro-organism often measured as an 'indicator organism' for the presence of other, more dangerous pathogens. See page 72.
E. coli	0/100 ml	A micro-organism often measured as an 'indicator organism' for the presence of other, more dangerous pathogens. See page 72.
pH	6.5-9.5	Monitored because of the influence that pH has on levels of dissolved metals (which may be harmful) and because of the risk of corrosion in plumbing systems. See page 89.
Turbidity	1 NTU (4 in Scotland)	A measure of solid content. Solids can cause aesthetic issues and corrosion, but can also be home to bacteria. See page 78.
Additional parameters monitored in Scotland		
Coliform bacteria	0/100 ml	A micro-organism often measured as an 'indicator organism' for the presence of other, more dangerous pathogens. See page 72.
Nitrate	50mg NO_3/l	Can cause acute toxicity. See page 91.
Odour	Qualitative	Particular odours can be indicators of specific contaminant types, in addition to it being important for acceptability. See page 91.
Taste	Qualitative	Particular tastes can be indicators of specific contaminant types, in addition to it being important for acceptability. See page 91.
Lead	25 μg/l until 2014, 10μg/l thereafter	Builds up in the body over time and can be extremely dangerous, causing brain damage. See page 90-91, and page 102.

Table A1. Parameters monitored in 'other supplies'.

Other supplies (Type B in Scotland): The parameters that will always be monitored are given in table A1. Other parameters may be included based on the findings of the risk assessment (for example if other local supplies are known to contain a particular contaminant, or if the land use in the catchment area suggests a likely problem).

Action in the event of water samples failing standards

If a water sample fails the standards, the local authority will investigate further and this may also cost you up to £100. In general they will try to help you in improving your supply without resorting to legal measures, but the degree of formality and urgency of their approach will vary according to what the parameter is and how serious the breach is.

Failures of microbiological parameters are taken particularly seriously given the risk of immediate illness. The local authority will take steps to ensure that all users of the supply are aware of the issue, the degree of the potential danger it poses, and give advice on how to minimise the danger. Remember that the LA is required by the Regulations to take action whenever they find the water to be unwholesome, even where there is no risk to health – so be nice to them.

Whilst all failures of the water quality standards will need to be investigated and rectified, there is provision within the Regulations for temporary departures from the required standards for some parameters that are not considered immediately dangerous to health. These derogations will be for fixed periods of time, and take the form of a less stringent (but still specific) limit on the parameter concerned. During this time, improvements must be made to the water supply to bring it up to standard.

The Regulations also require that if a temporary derogation is granted, users of the supply must be informed of it. Your Environmental Health Team will advise on this process, and many will grant informal derogations for parameters not immediately dangerous to health in preference to using the regulatory provision.

Differences between England, Scotland, Wales and Northern Ireland

The PWS Regulations vary across the UK. The major differences are as follows:

Scotland: The Regulations are generally more detailed and specific than the English version. *The key difference is that all private water supplies are included in the Regulations, and supplies to single dwellings are not exempt.* Additional differences include:

- 'Large supplies' are referred to as Type A in Scotland, with 'other supplies' referred to as Type B.
- An additional set of regulations applies which allows financial grants to be awarded to householders in order to help bring their supplies up to standard.
- The relationship between volume of water supplied and frequency of monitoring of Type A supplies in Scotland is different to that for large supplies in England.
- Details regarding authorisations of temporary departures from parameter standards and the exact measures to be taken in the event of water quality failures.
- There is an additional regulation requiring that where the water supply serves public or commercial purposes a prominent notice must be displayed informing people that the water supply is from a private source.

Wales: Identical to the English Regulations.

Northern Ireland: Very similar to England, except:

- The monitoring frequency for 'other private supplies' is stated as annually, but with a potential to reduce the frequency to every five years if the Department of the Environment deems this reasonable.
- The degree of compulsion on the regulators is lower. For single dwelling supplies they are not obliged to carry out RA if requested to, but instead are required to offer 'appropriate advice'.
- The records that the Department of the Environment are required to keep on private water supplies are only of monitored supplies, rather than every private water supply.
- The NI Regulations include detail of what the risk assessment should consist of.
- The service is free of charge.

Appendix 2 –

Checklist of actions on a private water supply

The checklist in table 3.3 can be used as a starting point to determine what work is needed on your water supply. An example of its use is given here; whilst the client declined to implement several of the recommendations, the measures taken could be expected to considerably reduce risk. Identifying details have been omitted.

	Comment
General record keeping	
Supply category under the Regulations	Single dwelling in Wales, not used for public or commercial purposes, so not categorised under PWS Regulations.
Contact details for people using the supply	
Contact details for any other parties (for example landowner if source does not belong to householder)	Landowner of the catchment includes the householder, farm X, and Forestry Commission.
Contact details of people supplied, and nature of their use	Single household, but the neighbours supply is from an immediately adjacent surface water supply from a similar catchment area, so failures in one supply could indicate likely issues with the next door supply. Client and neighbour both made aware of the benefits of information sharing on their water supplies.
Schematic of supply	None initially available. See figure A2.1.
Description of the source, photos where relevant, grid reference	None initially available. Source is a stream on a steeply sloping north east facing hillside. This stream arises from a spring at approximately 250m altitude in an area used for sheep farming. Evidence of sheep droppings and cast sheep wool in area of spring. Shortly after arising, the stream flows into an area of conifer forestry plantation (owned by Forestry Commission Wales). The inlet point to the water supply is located at (grid reference), immediately above a forestry road. The inlet consists of a 25mm blue MDPE pipe with a 250 ml fizzy pop bottle covering the end with holes drilled in it. This pipe was contained within a 450mm clay pipe buried vertically in the stream, at a point where the stream forms a pool. This clay pipe structure would protect the supply inlet pipe from the full force of the flow.
Estimated daily volume supplied	Approximately 300 litres/day

Table A2(a). Example of how the checklist from Chapter Three (table 3.3) was used for a specific household.

	Comment
General record keeping	
Details of any treatment systems	Y strainer in kitchen/dining room where supply enters the house. Inspected, in good condition, no rust. String particle filter immediately after Y strainer. Inspected, some material on surface, but not excessive. Specification unknown. Brita Aluna jug filter for drinking water in kitchen. Client resistant to the idea of UV treatment for micro-organism removal, or any treatment involving chemicals. Recommended replacement with ceramic under sink filter unit.
Any test results applying to the supply	None reported. Recommended testing, but client declined.
Dates and details of any investigations	Original request for site visit was based on the fact that the inlet pipe blocked periodically and required pressure jetting to rectify. Given the emergency/immediate nature of this, charges for this service were high. The contractor undertaking this work had not installed a robust inlet strainer, opting to simply replace the fizzy pop bottle with another one. He was recommending a concrete skirt and cover around the inlet, at considerable cost. Client was not keen on this solution, and was looking for alternatives.
Details of any maintenance undertaken	Y strainer checked periodically (client unable to be specific about frequency). Suggested checking Y strainer on a monthly basis and modifying frequency based on experience. String particle filter. Replaced annually. Jug filter changed according to manufacturers specifications.

Table A2(b).

	Comment
Site issues	
Stagnant water or poor drainage around source	Given the structure surrounding the inlet point, stagnation is unlikely to occur unless the supply dries up. The supply has been sufficient in the 5 years the current occupier has lived there, and neighbours have not reported any insufficiency (>20 year period). Neither neighbour nor client have experienced any taste changes in the water (which can be an indicator of stagnation). Client informed of the need to check for stagnant water issues after long periods of dry weather.
Livestock access, wild animals	Inlet point is from a moderately well fenced area of FC land, so livestock issues in the immediate vicinity are sporadic and low level rather than constant. A 50 metre section of the upper water course runs through upland sheep grazing areas. Recommended that the integrity of the fencing should be checked annually.
Vulnerability to agricultural runoff or other farm wastes	Initially unknown. Farmer confirms that muck spreading does not occur.
Forestry activities	Possible. Checked with FC, and thinning, clear felling and other major works are not scheduled within next 10 years.
Awareness of private supply by those who may affect it	Not initially. Relevant farmer and Forestry Commission both made aware.
Waste disposal in area	None.
Industrial activities	None.
Presence of off mains sanitation	Yes, but all at least 30 vertical metres below water intake.
Other groundwater access points	Initially unknown. None within catchment seemed likely (based on local OS map, consultation with LA Private Water Supply Team and queries with neighbours).

Table A2(c).

	Comment
Supply specific issues	
Stock proof fence and interceptor ditches around source	N/A
Concrete apron around well/borehole	N/A
Cover/access point	Lid for storage tank was tight fitting and watertight. Client didn't feel that having a lockable cover was necessary.
General condition of supply	Inlet structure (clay pipe) secure. Storage tank in good condition for its age. Fencing around storage tank in need of repair. Recommended minor repairs, but client declined.
General condition of pipework	All buried, and no reports of freezing, so depth assumed sufficient. Pipework was blue MDPE where visible.
Vermin protection of overflows	None present on storage tank. Installed.
Coarse filter on surface water supply inlets	Insufficient (plastic drinks bottle over end of pipe). Replaced with stainless steel inlet strainer on MDPE, with additional coarse mesh in inlet structure.
Protection of other components	N/A
Backflow protection	None needed.

Table A2(d).

	Comment
Supply specific issues	
Nature of supply within the house	Plumbing system within house had been replaced 5 years ago. Header tank in loft insulated and covered. Some evidence of birds/bats/rodents in loft. No evidence of porcelain staining (for example from metal contamination), and no reports of discolouration of other water using appliances. Kitchen tap off rising main, but header tank feeding all other appliances. Client aware of the importance not to drink water except from jug filter.
Supply variability (quantity or quality)	Supply not reported to be insufficient in summer. No taste, colour or odour issues reported.
Treatment system specific issues	Treatment consisted of Y strainer, string filter, with additional jug filter for drinking water. All of these appeared to be being maintained reasonably. See above.
Other comments/summary	The client had a very specific issue with supply failure due to pipe blockage and was not keen on much other work being done on the supply. The client was unconcerned about water quality. This is an extremely common occurrence. Giving a limited amount of written advice in these areas, providing a diagram of the supply, and detailing the nature of the supply and action to be taken in the event of changes in quantity or quality may be of use in the future by other contractors working or the supply, or future occupants of the house.

Table A2(e).

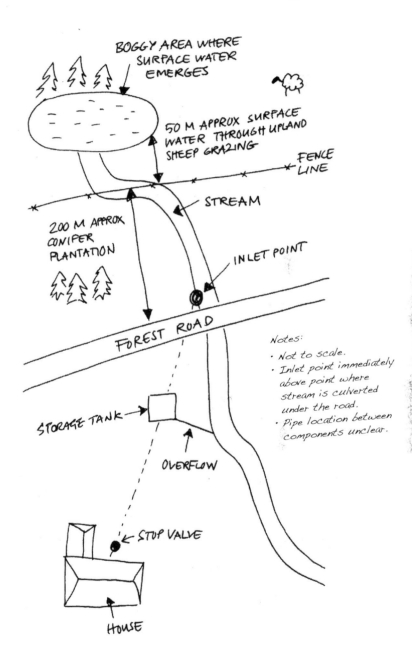

Figure A2.1. Diagram of the layout of the private water supply referred to in the case study above.

Acknowledgements

Judith is grateful for the contributions made by former-colleagues at CAT, particularly fellow biologists and the team in the publications department. Thanks are also due to those who reviewed chapters in earlier versions, including Chris Hartley, Chris Laughton, Peter Harper, Nick Grant and Cath Hassell. I am grateful to David Clapham who provided useful guidance on recent regulatory changes and perspective from years of work as an Environmental Health professional. Thanks are also due to current colleagues at the University of Leeds, in particular Martin Tillotson.

The following individuals and organisations kindly provided material for the illustrations:

Access Irrigation
Amos pumps
Avonsoft
Building Research Establishment
DroughtBuster
Environment Agency
Fairey Ceramics
Future Water Filters
Green Building Store
Green Shop
Hansa
Meterological Office
Metro Home Inspectors
Opella
Practical Action, formerly known as ITDG
Royal Doulton
The Water Label Company
Water Two

Glossary

Adsorption – the binding of molecules to a surface (for example as occurs in a carbon filter).

Alkaline – of high pH (the opposite of acid).

Anion – a negatively charge ion (e.g. Cl^-).

Aquifer – an underground area that can act as a water store (for example sand, limestone).

Artesian well – a well whose water level is higher than that of the surrounding ground water owing to a high hydraulic pressure within the aquifer.

Bladder storage tank – a type of pressure storage tank containing a water filled bladder.

Cation – a positively charged ion (for example Na^+).

Coagulation – the clumping together of fine particles into larger particles (which are then easier to separate from the water).

Coliform – a type of bacteria that is often used as an indicator organism; a sign that dangerous micro-organisms may be present.

Colloid – a suspension of solids in water. The solids tend to have negative surface charges so repel each other and are evenly dispersed in the water.

Diaphragm storage tank – a type of pressure storage tank in which a rubber membrane separates the gas filled compartment from the water filled compartment.

Environmental Footprint Analysis – a technique used to measure the environmental impact of a product or activity, based on the amount of land area required to support that product or activity.

Flocculation – a method to remove the suspended solids from water by causing them to stick together and therefore sink faster.

Flow regulator – a valve designed to limit the amount of water that can flow through a pipe.

Grey water – waste water from household sources with the exception of toilet water, for example showers, washbasins, sinks, washing machine, dishwasher.

Life Cycle Assessment – a process by which the environmental impact of products can be compared, by adding up their environmental impacts during manufacture, use and disposal.

Pathogen – a micro-organism that may be dangerous to health.

Perched aquifer – an isolated area of groundwater with a small recharge zone, rarely suitable for a private water supply owing to it's likelihood to run dry.

pH – a logarithmic scale of acidity. High pH is alkali, low pH is acid.

Pressure storage tank – a water storage tank that is designed to minimise pump cycling by allowing water to fill and empty over a range of pressures, similar to a pressure expansion vessel in a sealed heating system.

Protozoa – a type of micro-organism, often living parasitically (for example Giardia, Cryptosporidium).

Rapid gravity sand filter – a water purification technique which removes solids from large volumes of water relatively quickly, with the disadvantage that energy is required for frequent backwashing to remove the solids from the filter.

Residual disinfection – the continued presence in water of a compound that can kill pathogens (for example chlorine).

Reverse osmosis – a water purification technique in which water is passed under pressure through a membrane with very small holes in it.

Slow sand filter – a water purification technique in which water flows by gravity through sand, which acts as a filter to remove solid particles and micro-organisms.

Specific heat capacity – the amount of heat needed to raise the temperature of water.

Submersible pump – a pump that operates under water by pushing the water through the delivery pipe.

Suction pump – a pump that operates above water by sucking water into itself.

Transpiration – process by which plants lose water to the air through the pores in their leaves.

Upflow sand filter – a water purification technique in which water is pumped upwards through sand to remove solids.

Virus – a disease causing micro-organism without a nucleus.

Acronyms

DBP – disinfection byproduct

DWI – Drinking Water Inspectorate

EA – Environment Agency

EH – Environmental Health

EHS – Environment and Heritage Service (the EA equivalent in Northern Ireland)

MDPE – medium density polyethylene (a type of plastic pipe)

NSF – National Science Foundation (in America)

RO – Reverse Osmosis

RWH – rain water harvesting

SEPA – Scottish Environmental Protection Agency

UKAS – United Kingdom Accreditation Service

WHO – World Health Organisation

WRAS – Water Regulations Advisory Service

WRc – Water Research Centre

Index

Abstraction licence: 42.

AC current: 124.

Acid: 51, 89-90, 92, 97, 104, 179, 209-210.

Acidic: 37, 59, 91, 142.

Activated carbon: 94.

Activated carbon filters: 94-95, 138.

Additional parameters: 194-195.

Adsorption: 80, 94, 101, 209.

Aeration: 92, 173.

Air-gap: 139-140, 147-149.

Air release valve: 130.

Air release valves: 130.

Algae: 43, 54, 78, 91, 110.

Alkali: 89, 179, 210.

Aluminium: 70, 78, 93.

Alzheimer's: 93.

Ammonium: 70, 91-92.

Antimony: 70.

Aquiclude: 38.

Aquifer: 38-39, 43-44, 48-50, 55, 209, 210.

Aquifer (perched): 38, 210.

Aquitards: 37, 39.

Artesian well: 38-39, 209.

Asbestos: 142.

Atmosphere: 4, 8, 104.

Atmosphere: 6.

Atmospheric water: 6.

Backflow: 56, 67, 101, 129, 147-148, 159, 165, 187, 203.

Backwash: 68, 179.

Ball valve: 22, 80, 127, 146.

Bath: 17, 26, 30, 61, 111, 180.

Bicarbonate: 97.

Bilge pump: 123.

Bleach: 134.

Blue baby syndrome: 91.

Boiling: 86, 89, 96.

Borehole: 33, 39, 40, 42, 44, 48-49, 51, 56-57, 67, 92, 119-120, 123-124, 126-128, 134, 161, 188, 203.

Bromine: 87-88, 164.

Buffer: 41, 45, 49, 54, 109, 111.

Cadmium: 70, 90.

Calcium: 7, 97, 104, 134.

Carbon filter: 79-80, 89, 92, 94-95, 102, 138, 209.

Category: 64-65, 147, 193, 200.

Centrifugal pumps: 126.

Ceramic candle: 83-84, 96.

Ceramic filter: 79-80, 83-86, 93, 96-97, 99, 100.

Check valve: 101, 184, 187.

Chloramines: 88, 92, 104.

Chloride: 70-71, 90-91, 97.

Chlorine: 73, 86-89, 92, 95-97, 99, 102, 104, 134, 164, 210.

Chlorine dioxide: 88.
Cistern displacement device: 23-24.
Class: 156.
Clinometer: 113.
Closed loop: 1, 161, 167.
Coagulation: 12, 78, 96, 209.
Coliforms: 62, 69-70, 74, 85, 92, 93, 142, 148, 163, 166, 195, 209.
Colony counts: 70.
Combined sewer: 35.
Compost toilet: 3, 29.
Compost toilet: 25, 138, 161, 163, 166, 179.
Conductivity: 41, 69-70, 90, 195.
Cone of depression: 44.
Copper: 70, 102, 115-117.
Corrosion: 90, 97, 116, 154, 195.
Cryptosporidium: 73, 86, 88-89, 210.
Cyanide: 70.

DBPs: 88, 104.
DB valves: 184.
DC current: 124.
Deep well: 124.
Diaphragm: 123-125, 209.
Diarrhoea: 72, 74.
Dipper: 50.
Directly pumped: 141.
Directly pumped: 139-141.
Dishwasher: 17, 20, 101, 210.
Dishwasher: 17, 163.
Disinfection: 12, 73, 86-89, 96, 102, 104, 144, 148, 165-166, 175, 210-211.
Disinfection: 165.
Dissolved materials: 5, 77-78, 94, 97, 99.

Dissolved substances: 5, 42-43, 82, 89, 91, 94, 106, 144.
Distillation: 89, 96.
Dolomite: 68, 90, 93, 97.
Downpipe: 137, 144-145, 153, 176, 178.
Dowsing: 41-42.
Dowsing rods: 41.
Drainage: 14, 33, 35, 43, 54, 66, 116, 118, 143, 166-167, 171, 175, 191, 202.
Draw down: 44.
Dynamic pressure: 114-115.

Ecological footprint: 191.
Electronic pump controller: 126.
Environment Agency: 12-13, 15-16, 19, 31-32, 34, 40, 42, 49, 54, 57, 155, 158, 207, 211.
Evaporation: 8, 143, 169, 170-171, 173-174, 176, 182-184.

Faecal coliforms: 142.
Faeces: 7, 47, 55, 72-74, 144.
Filter: 35, 53-54, 67-68, 77, 79-86, 88-90, 92-97, 99-101, 103, 106, 122, 127-128, 132, 138-140, 144, 145-146, 148-150, 152-154, 166, 178-179, 182, 201, 203-204, 209-211.
Filter coefficient: 149-150, 152.
First flush: 144.
Float switch: 125, 127-128, 153.
Float switch: 125, 128.
Float valve: 23-24, 118, 129, 153.
Flocculation: 12, 78, 96, 209.
Flow rate: 17, 20-21, 25-26, 45-46, 68, 79, 82, 84, 86, 88, 112, 114, 116, 118, 150, 174, 178.

Flow restrictor: 174.
Flow restrictors: 26.
Fluoride: 70, 103.
Foot valve: 125, 128-129.
Fossil water: 4.
Friction: 115.
Fungi: 72.

Gastroenteritis: 72.
Giardia: 73, 86, 89, 210.
Granular activated carbon: 94.
Gravity fed: 29, 132, 139-141, 146.
Grey water: 5, 3, 33, 61, 88, 96, 161-169, 171, 174-175, 177-185, 187-189, 210.
Groundwater: 4, 6, 11, 30, 35, 37-46, 48-49, 55-56, 66, 90-92, 94, 104, 111, 126, 131, 134, 137, 202, 210.
Gutter: 137-138, 142-144, 154, 158.

Hand pump: 123-124.
Hardness: 7, 61, 77, 104.
Hard water: 84, 87, 104-105.
Head: 48, 112-116. 118-122, 124, 129, 135, 176, 178.
Head loss: 114.
Helminths: 72.
Hippos: 23.
Hose-pipe: 185.
Hot water: 12, 14-15, 20, 28-29, 31, 112, 116, 131, 154, 191.
Hydraulic pressure: 38, 46, 209.
Hydraulic ram: 109, 121-122, 128-129.
Hydrogen ions: 89.
Hydrological cycle: 1, 2, 4-6, 33-35, 37, 40, 110, 161.

Ice caps: 4, 6.
Impeller: 120.
Indicator organisms: 72-74.
Indirectly pumped: 139, 141.
Infiltration gallery: 52, 126, 128.
Infusion pumps: 88.
Inlet structure: 79, 96, 109, 126, 135, 203.
Inlet valve: 23.
Interceptor ditch: 45, 47.
Iodine: 87-88.
Ion exchange: 68, 91, 95, 97, 99.
Iron: 70, 78, 86, 90, 92-93, 97.
Irrigation: 162, 168-169, 174-175, 177, 179-180, 182-189, 207.

Jet pumps: 120.

Lake: 4, 6, 37, 52, 54-55.
Land drains: 40-41.
Laser level: 113.
Laundry detergents: 179.
Lead: 69-71, 90-91, 101-102, 195.
Leak: 22-23, 130, 154.
Legionella: 29, 61, 62.
Level sensors: 129, 154.
Limestone: 37-38, 90, 209.

Magnesium: 7, 90, 97, 104.
Magnesium oxide: 90.
Manganese: 71, 90, 92-93, 97.
MDPE: 115-118, 200, 203, 211.
Membrane: 88, 95, 97, 99, 118, 164, 209, 210.
Mercury: 71, 87.
Micro-organism: 7, 37-39, 43, 72, 73-74, 77-79, 81, 83-89, 94-96, 99, 101, 110, 122, 142, 195, 201, 209-211.

Mould: 91.
Mulch: 170-172, 183, 185-186.
Multi-stage pump: 120.

Nanofiltration: 95.
Nitrate: 39, 69, 71, 78, 91-92, 97, 157, 195.
Non-potable: 35, 99, 111, 137, 161.
Non-return valve: 129.
Odour: 91-92, 97, 130, 164, 195, 204.
Overflow: 22, 26, 35, 46-47, 67, 101, 110-111, 120, 129, 131, 148-149, 153, 165, 167, 176.
Overflow pipe: 47, 101, 129, 131, 153, 176.
Oxidation: 88, 92.
Ozone: 88-89, 96.
Pathogen: 12, 34, 54, 72, 74, 81-82, 85, 106, 109, 133-134, 144-145, 147, 162-163, 176, 195, 210.
Perched aquifer: 38-39.
Percussion taps: 26.
Peristaltic pumps: 120.
PH: 59, 61, 68-69, 71, 77, 86, 87-93, 103, 178-179, 195, 209-210.
Phosphorous: 8, 178-179.
Photovoltaic: 124.
Physical conditioners: 106.
Piston pump: 119, 123.
Polybutylene: 116.
Polyester filters: 82.
Polythene: 116, 183.
Potable: 11-12, 35, 55, 60-61, 96, 97, 99, 109, 110-111, 116-117, 132, 137, 145,-147, 155, 157, 161, 165-166.
Potassium: 90.
Power shower: 17-18, 116.
Pre-filter: 82-83, 96, 97.

Pressure: 20, 25-29, 37-38, 46-47, 51, 54, 60, 62, 82, 84, 93, 97, 99, 101, 110, 112, 114-117, 119, 121-123, 125-128, 135, 141, 169, 175-176, 184-187, 201, 209-210.
Pressure regulation: 28.
Pressure switch: 125.
Priming: 119, 124-125, 129.
Protozoa: 72-73, 210.
Pumps: 5, 30, 44-47, 49-50, 88, 93, 97, 101, 109, 111-112, 115, 118-130, 132, 135-136, 138-139, 141, 145-146, 153-156, 167, 187, 207, 210.
Pump test: 44.
Rainwater: 5, 18, 29, 33-35, 37, 133-134, 137-149, 151-159, 161, 166-169, 173-178, 180, 183-184, 186-189, 191.
Rainwater harvesting: 5, 137, 139, 141, 143, 145, 147, 149, 151, 153, 155, 157-159.
Rapid gravity sand filter: 80, 210.
Recharge: 38, 43-44, 50, 104, 131, 210.
Reservoir: 6, 14, 37, 52, 54-55, 78, 84, 93, 111, 128, 169.
Residual disinfection: 86-87, 89, 96, 102, 210.
Resin: 68, 91, 93, 95, 97, 104.
Reverse osmosis: 79, 95, 97, 99, 210.
Rivers: 4, 6, 8, 11, 37-38, 40, 51-53, 81, 126, 128, 176.

Saline intrusion: 90.
Sampling: 40, 59, 63-65, 69, 72, 73-75, 77, 130, 193.
Sand filter: 54, 68, 79, 80, 81-82, 85, 90, 93, 100-101, 122, 128, 138, 166, 210, 211.

Sava flush: 23.
Scale inhibitors: 106.
Schmutzdecke: 81.
Sea water: 6, 37, 60, 97.
Septic tank: 38, 43.
Settlement: 54, 78, 96, 109-110, 111, 117, 164.
Sewage: 3, 9, 11-12, 14, 21, 33-35, 38-39, 91-92, 118, 154, 156-157, 162, 164, 166, 168, 176, 179, 180, 191.
Shallow well: 48, 123.
Shower: 14, 17, 18, 20, 21-22, 25-26, 30, 111-112, 116, 133, 143, 144, 161-164, 177, 180, 182, 188, 191, 210.
Silver: 83, 85, 88, 96.
Sine rule: 113.
Siphon: 23-24, 180, 187.
Siphon flush: 23.
Soakaway: 35.
Soaker hose: 183.
Sodium: 71, 178.
Sodium hypochlorite: 134.
Soft water: 91, 97.
Solids: 5, 12, 14, 42, 51-52, 54, 77, 78-81, 83-85, 89, 93, 96, 101, 106, 109, 124, 130, 133, 143, 146, 164, 176-179, 181-182, 195, 209, 210-211.
Solvent: 7-8, 97, 169.
Source control: 9, 60, 63-64, 74, 77, 85, 106, 130, 194.
Spray fitting: 26.
Spring: 33, 38-39, 41, 44-48, 67, 82, 92, 111, 131-132, 134, 161, 171, 200.
Spring box: 47.
Sprinkler: 62, 169, 174-175, 182, 183, 188.
Static pressure: 114-115.
Steel coil filter: 79, 82.

Stop tap: 17, 103, 110-111, 129.
Strainer: 47, 51, 52, 54, 111, 124, 128, 130, 201, 203-204.
Stream: 4, 35, 37, 39-41, 44, 51-55, 74, 93, 121-122, 128-129, 133, 162, 178, 200, 205.
Submersible pump: 119, 124, 125-126, 146, 210.
Suction pump: 119, 123-129, 210.
Sulphate: 71, 78, 91, 97.
Supply pipe: 3, 14, 17, 81, 116, 118, 119, 121, 125, 129, 147, 187.
Surface drip: 183.
Surface water: 13, 30, 33-37, 40, 41-45, 47-48, 52, 54-56, 66-67, 73, 79, 90-92, 96, 100, 104, 110, 118, 121, 126, 128, 131-134, 137, 143, 157, 161, 167, 171, 173-175, 191, 200, 203.
Surge tank: 182, 186.
Swimming pool: 62, 96, 134, 188.

Tap: 3, 8, 9, 11-12, 17-18, 20, 22, 26-29, 61, 67, 74, 101- 103, 106, 110-111, 114, 116-118, 129, 147, 176-178, 184, 186-188, 204.
Taste: 4, 7, 8, 60-61, 71, 88-89, 91 -92, 95-97, 101-102, 104, 130, 195, 202, 204.
Toilet: 3, 14, 17-18, 20, 22-25, 29, 31, 36, 59-62, 99, 129-130, 133, 137-139, 144, 146-148, 152, 154-155, 157-158, 161-167, 177, 179-180, 191, 210.
Total organic carbon: 71.
Transpiration: 4, 9, 169-170, 211.
Trickle pipe: 183.
Turbidity: 68-69, 71, 78, 81-82, 96, 195.
Ultraviolet: 43, 54, 68, 84, 96, 117.

Urine: 7, 20, 162.
UV: 54, 81-82, 85-87, 96, 100, 101, 106, 110, 122, 127-128, 132, 148, 201.

Vacuum flush: 25.
Valve flush: 23.
Variable flush lever: 23-24.
Vermin: 46, 67, 110-111, 203.
Vermin-proof overflow: 111.
Virus: 72, 86, 88, 211.

Washbasin: 143, 162, 180-181, 188, 210.
Washing machine: 14-15, 17, 20, 28, 30, 61, 101, 137, 141, 144, 152-153, 162, 163, 175, 177, 182, 187, 210.
Water audit: 17.
Water brake: 28.
Water butt: 137, 171, 176-177, 187.
Water cycle: 175.
Water efficiency: 3, 11, 15, 17, 18, 21, 26, 28-34, 36, 114, 132-133, 154-156, 158, 161, 166, 168-169, 173, 177, 191.
Water meter: 14-16, 21, 29, 130, 156.
Water pressure: 20, 26-29, 47, 54, 84, 93, 97, 101, 114, 116, 125, 127, 128, 141, 176, 184.
Water softener: 104, 179.
Water test: 43, 75, 91, 106, 117, 192.
Water timer: 186.
Well: 40, 44, 48-52, 56-57, 67, 92, 120, 123.
Wind pump: 109, 120-121, 136.
WRAS: 100, 109, 135, 147, 158-159, 187, 211.
Zeolite: 179.